TRANSPORT OF HAZARDOUS MATERIALS

Proceedings of the symposium held in London
on 15 1977

The Institution of Civil Engineers
London, 1978

Symposium sponsored by the Council of Engineering Institutions and the Council for Science and Technology Institutes.

Main Organizing Committee

Sir Norman Rowntree (Chairman), J. H. G. Chinnock, R. Main, Dr F. S. Feates, B. Hooper, Dr E. G. West, Dr D. Train, H. G. Riddlestone, C. D. Cernes, A. D. Smith, R. E. Bloomfield, J. B. Story, G. A. King and Dr M. E. Peover.

Production Editor: Mary Monro

ISBN 0 7277 00588

© ICE 1977, 1978

All rights reserved. Except for fair copying no part of this publication may be reproduced, stored in a retrieval system or transmitted in any form or by any means electronic, mechanical, photocopying, recording or otherwise, without the prior written permission of the Institution of Civil Engineers.

The Institution of Civil Engineers as a body does not accept responsibility for the statements made or the opinions expressed in the following pages.

Produced and distributed by Thomas Telford Limited for the Institution of Civil Engineers, PO Box 101, 26-34 Old Street, London EC1P 1JH.

Printed in Great Britain by The Burlington Press (Cambridge) Ltd. Foxton, Royston, Herts.

Foreword

Dr. E. G. WEST, OBE
Chairman of Working Party of the Organizing Committee

The problems arising from the transportation of hazardous materials, as outlined in the papers prepared for this symposium were further illustrated by the presentation of films and slides, as well as by an exhibition of photographs, at a reception on the evening before the meeting. The Organizing Committee and Working Group are grateful to all those who helped to obtain these visual aids, to the Authors for their expert contributions and to the opening speakers who set the style for the excellent discussions which are recorded in this volume.

In the UK there are several hundred incidents a year involving the spillage of dangerous loads, and the examples which are described well illustrate the wide range of materials which can prove hazardous. That so few accidents escalated to disasters provides striking proof that much has already been accomplished, but with the increase in the volume of traffic and in the types of load there is a continuing need for the assessment of risks followed by further studies on their reduction.

Petrol tankers are the most numerous and familiar vehicles conveying flammable materials and are, of course, regarded as an acceptable risk in spite of occasional accidents. The increasing conveyance of liquid fuel gases at cryogenic temperatures imposes additional problems, due to the effects of very low temperature on most steels and other materials, the different requirements of the emergency services and the possible need for special traffic regulations on routing. At the other end of the temperature scale, liquid metals, particularly aluminium, may be carried in special ladles at over 800°C from refineries to foundries, for which a Code of Practice has been successfully operated for a number of years. Loads of materials which may release clouds of gas, such as chlorine or ammonia, demand rapid emergency action based upon information which must be immediately available. Many powdered substances in bulk can prove dangerous, whilst the same materials are harmless in small containers. Mixed loads can be particularly hazardous and a wide variety of goods such as paint in small quantities, are inocuous until they are involved in serious accidents.

Chemical manufacturers and users are well aware of the properties of the substances which they handle and they have developed schemes for identification and action when accidents occur. It is more difficult to deal with situations which may arise when waste materials are being transported but even here good practice can reduce possible risks significantly.

FOREWORD

Clearly there are major considerations of communication involved in all incidents which require the emergency services to work closely with the scientific expert and the technologists responsible for long-term as well as immediate action. There was much valuable exchange of information and discussion between all parties during the symposium.

The Organizing Committee prepared a programme which was intended to bring together representatives of wide professional interests, and the discussions indicate that the scope of the papers had met their requirements. Much remains to be done in the fields of risk assessment, transfer of information between the different professions concerned, drafting of regulations, standards and codes of practice and the investigation of failures. The integration of such activities on an interdisciplinary basis is necessary, as noted in one of the papers which says '... The effects of a large-scale incident in a public place are so horrendous that no conscience should be stilled by the difficulty of finding a workable solution'.

CONTENTS

Opening address   1

1. The transport by road of hazardous substances: technical regulatory and international problems. H. J. DUNSTER   3

2. Schemes for the rapid identification of chemicals and their hazards in an emergency. F. R. CUMBERLAND and F. S. FEATES   9

Discussion: Papers 1 and 2   33

3. Design features for road vehicles. C. D. CERNES   49

4. Design of tank containers. C. O. FARMER   61

Discussion: Papers 3 and 4   73

5. Routing of hazardous substances moved by road. W. G. ASHTON   87

6. Monitoring, including control systems, codes of practice and some positive proposals for future action. E. J. WILSON   99

Discussion: Papers 5 and 6   115

7. Emergency response systems. A. H. SMITH   127

Discussion: Paper 7   137

8. General principles of risk assessment. F. R. FARMER   141

Discussion: Paper 8   147

Summing up   151

Opening address

Air Marshall Sir CHARLES PRINGLE, KBE, MA, FRAeS, Chairman
Council of Engineering Institutions.

This symposium is one of the first results of the collaboration between the Council of Engineering Institutions and the Council for Science and Technology Institutes. Through their 25 constituent bodies they represent scientists, engineers and other technologists totalling some 200 000 individuals. Nearly 6 years ago the Councils set up a joint Technical Committee to consider those subjects which include more than one discipline, with the aim of defining problems and encouraging action on them by discussion and other means.

The transportation by road of substances which can prove hazardous is clearly of interest to both engineers and scientists in a number of disciplines, especially because, as well as the technological aspects, they present important social and economic problems which require urgent attention. The increasing volume of road traffic, coupled with the increase in the number of potentially hazardous substances which need to be moved around Britain, has not gone unheeded by government, local authorities, the industries concerned, nor indeed the general public. Many of us are aware of incidents which have occurred and which might have caused major disasters.

We do not seek to alarm the general public, but accidents will inevitably continue to happen due to human failings. Engineers and scientists have a corporate responsibility therefore to seek to reduce the resulting dangers to human life. So the engineering and scientific professions must be alerted to the nature of the risks which may arise so that all can co-operate to reduce the chance of accidents and their consequences to life and property. The value of the work already undertaken by vehicle builders and users, the emergency services and government agencies, can be extended and enhanced through activities undertaken jointly or individually by the professional institutions. Mechanical and chemical engineers, chemists and physicists, metallurgists and welding experts, civil engineers and traffic specialists are all involved.

The papers and discussions will disseminate current information, define problems requiring further investigation and stimulate the attention of all the professions involved.

The scope of this symposium is limited to road vehicles although the hazardous substances with which it is concerned are, of course, carried by rail and by ship. Although the conditions may be different, many of the risks arise from the same cause, and there are other hazards when trans-shipment, loading and un-

OPENING ADDRESS

loading operations are in progress. However, the Committee decided that this conference should aim to produce for the majority a sobering unease and a realization of problems which are already being and will continue to be tackled by those scientists, engineers and others directly and indirectly concerned.

2

1. The transport by road of hazardous substances: technical, regulatory and international problems

H. J. DUNSTER, BSc, ARCS, Health and Safety Executive

Although the actual damage and injury caused by the road transport of hazardous substances is small, the potential for danger is substantial and public concern is real. The present legislation is incomplete and complex, and is out of line with European and international developments. The Health and Safety Executive is planning to base future legislation on a framework compatible with United Nations proposals and also to encourage the adoption of UN proposals by European authorities. This is an easily stated but not an easily achieved objective. Indeed, the objective itself may contain some unreconcilable conflict. Transport, like public relations, is a subject about which everyone thinks he is an expert. This conference will have served its purpose if it demonstrates the complexity of the problems.

The fact that this conference has been organized at all is a demonstration of the concern felt about the transport of hazardous substances, particularly transport by road. This development is an excellent example of what I think is a relatively modern trend: the tendency to be concerned, perhaps more concerned, about potentially dangerous events than about the actual and immediately dangerous events taking place all around us. At present about 7000 people a year are killed on the roads in the UK and perhaps five times that number are seriously injured. This arises partly for good economic reasons and partly because we wish to enhance our own convenience. A good deal of money is spent to prevent the numbers from being even higher but few people are really prepared to surrender any significant part of their personal freedom or standard of living in order to bring these numbers down by any substantial amount. Most deaths on the road occur in ones and twos, and it is very rare for a single accident to cause deaths running into two figures. One has to take into account the fact that the public quite genuinely worries more about a single event killing 10 or 20 people than about 10 or even 100 times more people being killed one at a time.

2. It is this possibility of a single accident killing and injuring many people that is at the root of the concern about the transport of hazardous substances. At present the record is extremely good, perhaps surprisingly good. Over the past few years there have been fewer than 2 deaths/year attributable to hazardous substances in transit and, so far as can be judged, the 3500 road tankers in

Transport of Hazardous Materials, ICE, London.

the UK are involved in some sort of incident concerning their load no more than once in about every 5 million miles travelled. However, this can only be an estimate and there must be many more incidents such as leakages from packages, drums falling from the back of lorries, leaks in hose couplings and so on, which are never reported to anyone.

3. Some of the incidents that are reported and achieve publicity are important in their own right. Thus a petrol tanker crashing and catching fire not only causes damage and injuries and possibly deaths at the scene of the accident; it also gives rise to widespread worries about what might have happened if the location of the crash had been more critical.

Current legislation

4. Largely because of the potential for danger, rather than because there is any evidence of a bad record, there has been some legislation to regulate the handling of a limited number of dangerous substances since 1881, when 'an Act to regulate the hawking of petroleum' was put on the statute book. Since then, with the increasing complexity of modern life, more and more substances with very wide ranging hazardous properties, many quite different from those of petroleum, are being conveyed in larger and larger quantities. This was recognnized by piecemeal changes in the law leading to the Petroleum (Consolidation) Act of 1928, which empowered local authorities to grant licenses to keep petroleum spirit. These authorities also have power within their licensing area to enforce regulations for the conveyance of petroleum and a number of named flammable liquids, corrosive substances and organic peroxides. The requirements are in very general terms and can be summarized under five headings:

   (a) petroleum spirit must be safely discharged into storage tanks;
   (b) all due precautions must be taken against spillage;
   (c) all due precautions must be taken against accident by fire or explosion;
   (d) some specified labelling requirements must be met;
   (e) any accident involving the dangerous substance and causing loss of life or personal injury must be forthwith reported to the Health and Safety Executive.

Additionally there is a more general requirement that any road tanker for the conveyance of petroleum spirit in quantities in excess of 1500 gal. must be constructed to an approved design.

DUNSTER

New developments

5. The need for improvement was recognized some while ago by the Home Office, which was at the time responsible for regulating the conveyance of hazardous substances by road, and work was started towards bringing in more general control. The main problems arose from the fact that the controls at that time were made under the Petroleum (Consolidation) Act 1928, and this requires that any hazardous substance to be dealt with by the Act has to be specifically named. There are already seventeen separate pieces of legislation under this Act, and effective cover over the bulk of the remainder is estimated to involve another 25 separate pieces of legislation. In addition to the obvious complexity of such a jungle of legislation there is the severe problem that information on the quantities transported and the details of the transport operations are simply not centrally recorded; the data are thus lacking for any logical approach to the regulation of transportation by a series of regulations dealing with separate materials.

6. A review of this situation led to the conclusion that it would be better to use the Health and Safety at Work Act as a basis for a more fundamental revision of the conveyance regulations.

7. The UK is a signatory to the agreements concerning the international carriage of dangerous goods by both road and rail, and both of these agreements make use of the United Nations system of classification although there are still differences between the details of the rail transport agreement and the UN system. In principle therefore, the Health and Safety Executive has decided to use the UN system of classification and to attempt as far as possible to ensure compatibility between our regulations based on the UN classification and those of the existing agreements.

8. The UN classification is descriptive and, at least in principle, a new substance can be slotted into the classification on the basis of its properties and characteristics. The classification includes such headings as explosives, oxidizing substances and corrosives; in practice therefore it is not always possible to allocate a new substance unequivocally to the list, and the final decision contains a small element of arbitrariness.

9. The UN maintains a committee of experts amongst whose many tasks is that of allocating substances to the UN classification. This committee has a difficult task, and at times has to face difficult issues of principle. For example, there is still debate as to whether a substance which may be harmful as a result of prolonged occupational exposure over a working lifetime should also be regarded as hazardous in the short-term situation following a transport accident.

10. Despite the fact that the UN views are recommendations and have been differently treated by the organizations concerned with different forms of trans-

port, there is now an increasing move towards a consensus based on these UN recommendations. The UK has been active in encouraging moves towards such a consensus, and while considerable progress has been made there is still much to achieve. One of the conflicts still not fully resolved relates to the use of 'open' or 'closed' classes. If open classes are used, materials not already named can be dealt with, by comparing them with existing substances, while closed classes means that a substance not named in the class cannot be carried on an international journey without an individually negotiated special agreement. At present some 1700 substances are identified in the UN list as being moved in substantial quantities on a worldwide scale, and while this may cover the more important examples it is probably no more than 10% of the total substances moved in various ways and in various quantities.

11. There is no obvious and simple solution to these complexities but it is hoped that the Health and Safety Commission will publish the general outline of proposed regulations soon and that these will be received with a sufficiently general approbation that the Commission can move on towards proposing regulations and the necessary supporting guidance reasonably soon after that.

12. Despite this time-scale the Health and Safety Executive felt it right to advise the Commission to issue a consultative document on its tanker marking scheme because this was widely felt to be a matter of urgency and because the detailed arrangements had already been tried out in an existing voluntary scheme. There should be no difficulty in incorporating a hazard information scheme into the general regulations when these follow.

13. Even in this simple area there is a divergence between the UK and our overseas colleagues. At the moment the UK scheme is alone in containing an action code which gives immediate information to the emergency services as to the steps they should take. Information about the properties of the material is available only through an identification code which has to be checked back with the emergency services' control base. The European and UN approach is to give immediate information on the properties of the substance so that emergency services can make their own decisions. In the Executive's view these proposals take insufficient account of the pressures on emergency services at the time of an accident and also underestimate the difficulties of providing enough information about properties in a coded form. In this respect the UK is also in a minority and the ultimate outcome is by no means clear yet.

Road versus rail

14. Having mentioned the work going on concerning road and rail transport, I might perhaps break off to say a word about the choice between road and rail for the conveyance of dangerous substances.

15. There is a widespread view that it would be safer to transport a higher proportion of hazardous substances by rail and thus get them off the roads where accidents appear to be more common and often more spectacular. This is based on the general view that personal travel by train is safer than personal travel by road and, if trains are compared with private cars, this is certainly so. The data are less convincing between trains and buses, and between trains and commercial transport generally and, in any event, the accident involving the vehicle itself is not the only factor to take into account. One other factor is certainly the geographical relationship between the mode of transport and population centres. In the UK we are slowly moving much of trunk road transport away from highly populated areas, while the railways still run right through the middle of heavily congested built-up areas. A second factor is that the use of rail transport is often possible for only a part of the journey. Many of the major chemical manufacturers have railheads, but many of their customers do not. If there is any retail distribution, then the customer certainly has no immediate access to a railhead. In such situations there is necessarily a transfer operation, followed by local movements by road. Would you be prepared to limit your purchases of petrol to out-of-town garages, or the nearest British Rail depot outside the town?

16. I do not want to suggest that the balance is necessarily right at present, but I would like to emphasize that the simple answer, 'send it by rail', is not necessarily either sensible or feasible.

The way ahead

17. There are several subjects on which every man is his own expert. Public relations, national economics, law and religion are well-known examples. It seems to me that transport is rapidly being added to the list. Everyone has views, and only too often it is not thought necessary to support these views by factual data. I have concentrated on the UN as a focal point and on the various agreements for the different modes of transportation as the basis for international agreement. I must also, however, refer to the European Economic Community and the interest in the so-called 'non-tariff barriers to trade'.

18. The subjects of transportation and labelling are closely related, and the European community has its own system of labelling which is different from any of those dealt with in the organizations I have mentioned. The EEC system is the subject of a directive which is binding on member states so, in this respect, freedom of action is limited to that which the UK can persuade its European colleagues to accept.

PAPER 1

19. I hope I have been able to show that there is no possibility of instant solutions adopted at a stroke. The Health and Safety Executive is operating in an international context and we would be doing this country a disservice if we allowed ourselves to be rushed into proposing domestic legislation which made life substantially more difficult for those concerned with international trade. There is a condition well known to medical men which is sometimes called *furor therapeuticus*; this describes the condition where the doctor feels obliged to take some allegedly therapeutic action regardless of his views on its true usefulness. Despite the pressures which will continue to be applied to the Health and Safety Executive and Health and Safety Commission to take urgent and decisive action in this field, I believe that the patient (in this case the British public) will be better served by a careful and thoughtful programme of legislation soundly based on wide consultation. Anyone can write bad regulations—getting this particular package right will call for more than usual care, understanding and patience.

2. Schemes for the rapid identification of chemicals and their hazards in an emergency

R. F. CUMBERLAND, LRIC and F. S. FEATES, BSc, PhD, FRIC,
Environmental Safety Group, Harwell

Accidents arising from the uncontrolled release of chemicals into the environment demand the immediate availability of essential information to enable appropriate countermeasures to be taken. The Paper outlines various schemes which have been introduced nationally and internationally to identify chemicals and their intrinsic properties (toxicity, flammability, etc.) in such situations. Means of overcoming language difficulties are also discussed. Computerized information retrieval systems are rapidly becoming a familiar feature in the field of environmental protection. The Authors describe some of these systems which are directly relevant to emergencies arising from the transport of hazardous materials.

Products of the chemical industry vary enormously in their type and degree of hazard, ranging from explosives through man-made fibres (which may produce cyanide on combustion) to plasterboard. The products are transported as solids, liquids or gases under a wide range of temperatures and pressures. The packaging methods employed will depend on the potential hazards of the product and the requirements imposed by regulations appropriate to the mode of transport. It is the purpose of this Paper to discuss various means by which potential hazards may be rapidly identified in the event of an incident involving chemicals or chemical products. Transport regulations, although closely related to vehicle marking schemes, are not considered in detail.

Hazard classification

2. The basic system of hazard classification now widely adopted for the transport of hazardous materials by land, sea or air, follows the recommendations of a United Nations committee of experts on the transport of dangerous goods.[1] Hazard types are divided into nine main classes represented numerically (1-9). Most classes are further subdivided into hazard divisions and subdivisions depending on appropriate criteria. Table 1 gives a simplified list of hazard types. The criteria adopted for classifying substances vary, depending on the appropriate regulations and mode of transport, e.g. the Intergovernmental Maritime Consultative Organization Dangerous Goods Code (IMCO),[2] the European Agreement concerning the International Carriage of Dangerous Goods by Road (ADR),[3] the International Regulations concerning the Carriage of Dangerous

Transport of Hazardous Materials. ICE, London.

Goods by Rail (RID),[4] and the International Air Transport Association Restricted Articles Regulations (IATA).[5]

Primary hazard identification

3. Each hazard class has an appropriate diamond-shaped pictorial label (Fig. 1). These labels are generally referred to as 'warning diamonds' and are intended for use in transport situations. Each label has a characteristic background colour:

Explosive	orange
Inflammable	red
Water reactive	blue
Oxidizer	yellow
Toxic/Infectious	white
Corrosive	black and white

Diamonds for radioactive substances have either a white or half yellow and half white background depending on the level of radioactivity. An additional diamond for other hazardous substances has recently been introduced in the UK as part of the proposed Hazardous Substances Tank Labelling Regulations 1977.[6] It consists of an exclamation mark on a white background.

4. The hazard warning diamonds may also have an approved inscription indicating the hazard and/or the UN class number, but the basic principle is that the symbol expresses the hazard pictorially and conveys a simple message ('diamond means danger'), thus overcoming language barriers. The value of an internationally accepted series of labels when clearly displayed on vehicles and packages is therefore obvious:

(a) they convey a non-specific warning to the general public;
(b) in an accident situation they provide an indication to the emergency services of the major hazard to be encountered.

Product identification

5. Hazardous chemical products are considered for classification by the group of rapporteurs of the UN Committee of Experts, based on information submitted by the manufacturer on standardized questionnaires (Appendix 1). If accepted, the product is added to the UN list and assigned a four digit substance identification (or UN) number.

Hazard information systems

6. In recent years, hazard classification in combination with a substance identification number (or chemical name) has formed the basis of hazard information systems in various countries. Emergency services and chemical producers

agreed that there was a need to ensure that essential basic information was immediately available in an emergency, irrespective of the mode of transport or whether on a domestic or international journey.

European system

7. The movement of dangerous goods by road across most international frontiers in Europe is subject to the ADR regulations already mentioned. These regulations require vehicles conveying scheduled chemicals to display an orange plate (Fig. 2) at the front and rear of the vehicle. Similar regulations (RID) apply to rail wagons except that the plates are fixed to either side. Each plate shows two identification numbers. In the lower part, a four digit UN number identifies the substance, whilst the upper section displays a two or three digit hazard identification number (previously referred to as the Kemler number). This hazard

Table 1. UN hazard classification

UN Class		Hazard type
1	1.1–1.5	*Explosives* Mass explosion hazard—very insensitive substance
2		*Gases* Compressed, liquified or dissolved under pressure
3	3.1–3.2	*Inflammable liquids* Flash point 23°C–23–60.5°C
4	4.1 4.2 4.3	*Inflammable solids* Spontaneously combustible substances Substances giving off inflammable gases in contact with water
5	5.1 5.2	*Oxidizing substances* other than organic peroxides Organic peroxides
6	6.1 6.2	*Poisonous (toxic) substances* Infectious substances
7		*Radioactive substances*
8		*Corrosive substances*
9		*Miscellaneous dangerous substances*

PAPER 2

Other hazardous substances
(excluding explosives and
radioactive substances)
(UK only)

Fig. 1 (*above and left*) Hazard warning diamonds

identification number is somewhat similar to the UN classification system, but the second and subsequent digits indicate secondary hazards. In the ADR/RID hazard identification system the first figure (primary hazard) is as follows:

 2 for gas   5 for oxidizing substance
 3 for inflammable liquid   6 for toxic substance
 4 for inflammable solid   8 for corrosive.

The second and third figures (secondary hazards) are:

 0 no meaning   6 toxic risk
 1 explosive risk   8 corrosive risk
 2 gas may be given off   9 risk of violent reaction from sponta-
 3 inflammable risk   neous decomposition or self poly-
 5 oxidizing risk   merization.

Where the first and second figures are the same, an intensification of the primary hazard is indicated, for instance, 33 means a highly inflammable liquid; 66 indicates a very dangerous toxic substance; 88 a very dangerous corrosive substance. When the first two figures are 22 a refrigerated gas is indicated. The combination 42 indicates a solid which may give off a gas on contact with water. Where the hazard identification number is preceded by the letter X this indicates an absolute prohibition of the application of water to the product.

8. Thus in the example given in Fig. 2 the information provided in numerical form indicates that:

(a) the substance is potassium metal;

(b) it is an inflammable solid which may give off a gas and possesses an inflammable risk;

(c) it must not be allowed to come into contact with water.

This European system provides information on the hazardous properties of the substance, overcoming language difficulties, but does not give any advice to the emergency services of what action to take or the type of protective clothing required. The UK is a signatory to the ADR and RID regulations, and in consequence vehicles exporting or importing scheduled chemicals to this country have to comply with this hazard information system.

UK Hazard Information System (UKHIS)

9. The UK Hazard Information System, for domestic traffic, was introduced as a voluntary tanker marking scheme (VTMS) in 1975 and is shortly to become mandatory. This system requires the display of an orange and black composite hazard information panel to the sides and rear of road tank vehicles and on

CUMBERLAND & FEATES

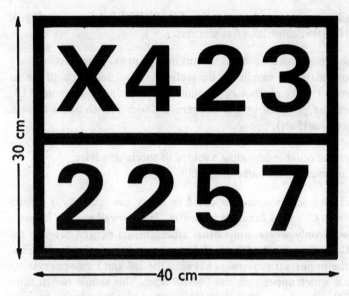

Fig. 2. ADR/RID hazard information panel

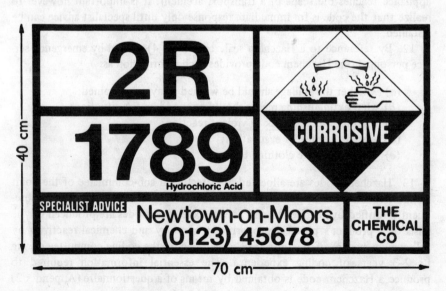

Fig. 3. UKHIS hazard information panel

either side of rail tank wagons. An example of a UKHIS label is given in Fig. 3.

10. The UKHIS label consists of five sections:

(*a*) an emergency action code (Hazchem) in the upper orange section;
(*b*) an appropriate UN number in the section immediately beneath the action code (in the absence of an allocated UN number temporary UK numbers are provided by the Dangerous Goods Branch of the Department of Transport);
(*c*) a hazard warning diamond;
(*d*) a telephone number providing a source of specialist advice;
(*e*) the company's house name or symbol.

The specialist advice telephone number could be either the company, a consortium or the Harwell Chemical Emergency Centre. The Harwell number however would be displayed only where contractual arrangements existed between the company and Harwell under the Chemical Emergency Agency Service Scheme.

11. One of the important differences between the UK and European systems is the action code which appears on the UKHIS label. This simple two or three character code (Hazchem) was devised by the London Fire Brigade and has now been adopted by central government and the chemical industry. It specifies the first aid action to be taken by the emergency services as soon as the first appliance reaches the scene of a transport accident. It is important however to realize that the code is for immediate response only until specialist advice can be obtained.

12. By reference to a Hazchem scale card (Fig. 4) carried by emergency service personnel, the Hazchem code provides such information as:

(*a*) whether the spillage should be washed away or contained;
(*b*) what extinguishing media should be used in a fire situation;
(*c*) whether there is a risk of violent reaction;
(*d*) a need to consider evacuation;
(*e*) what protective clothing to wear.

13. Hazchem codes are allocated by a technical sub-committee of the Joint Committee on Fire Brigade Operations within the Home Office, and subsequently ratified by the Health and Safety Executive. Codes are produced on the basis of a product's physical properties, toxicology and chemical reactivity by following a system of 'decision trees' developed by the coding committee during its 2–3 years of coding experience. The essential information required to produce a Hazchem code is obtained by means of a questionnaire (Appendix 2) completed by the organization requesting the code.

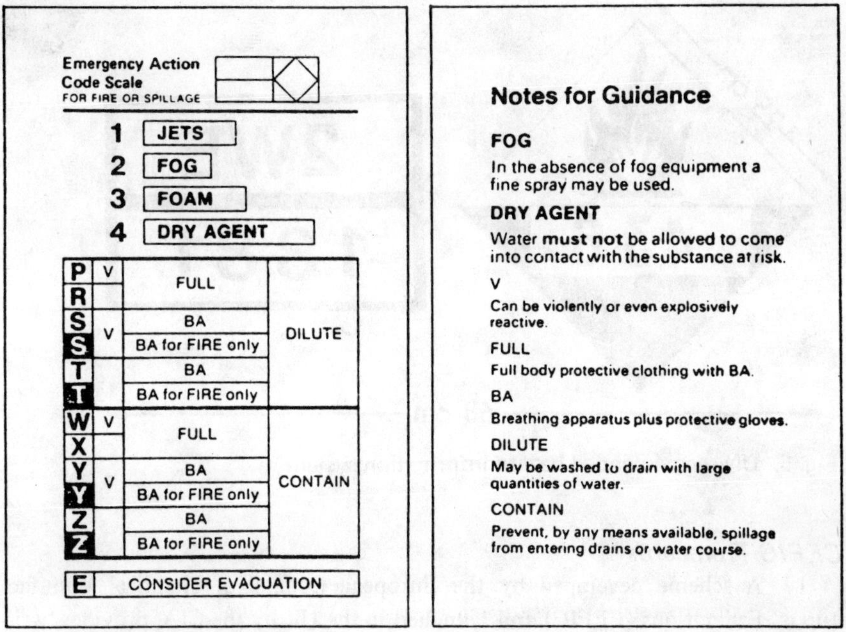

Fig. 4. Hazchem scale card

14. Lists of approved Hazchem codes and UN numbers/chemical names are issued to the emergency services by the Home Office. The Chemical Industries Association (CIA) publishes similar information[7-9] for other organizations.

15. The European hazard identification (Kemler) number is therefore a 'properties' code whereas the UKHIS (Hazchem) classification is an action code. Each has its particular merits and there has been considerable international discussion aimed at bringing the two schemes together in a unified system. The UN committee of experts meeting in Geneva in December 1976 recommended that a hazard information system, combining action and properties codes, be used in all modes of transport throughout the world. This new system (Fig. 5) includes a conventional hazard diamond in which the properties code is displayed. Alongside the diamond would be an orange plate bearing an action code (upper section) and the UN number. There is, however, no indication when such a scheme may come into operation.

Transport emergency cards

16. As well as labelling vehicles it is advisable for the driver to carry more detailed instructions in his cab with a cargo manifest. Typical standardized information sheets include Tremcards and Chemcards.

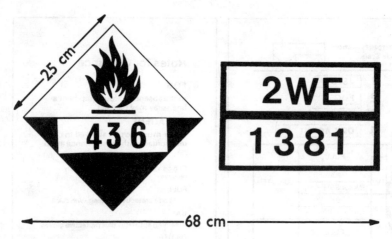

Fig. 5. UN recommended hazard information system

CEFIC Tremcards

17. A scheme developed by the European Council of Chemical Manufacturers' Federations (CEFIC) and launched in the UK by the CIA, provides 'written instructions' in the form of standard transport emergency cards (Tremcards). These cards (Fig. 6) have a standard format, A4 size, and use standard phrases agreed internationally by the fourteen member countries of CEFIC (Austria, Belgium, Denmark, Eire, Finland, France, West Germany, Great Britain, Holland, Italy, Norway, Spain, Sweden and Switzerland).

18. The initial purpose of Tremcards was to assist suppliers of chemicals throughout Western Europe to comply with the 'written instructions' requirement of ADR. This requires that drivers of road vehicles carrying scheduled chemicals between or through countries which are parties to ADR must carry instructions in writing for each substance in the language of the country of origin and the languages of countries of transit and destination.

19. A further purpose of the CEFIC scheme was to assist suppliers to comply with any domestic legislation within a particular country, though no such requirements apply at present in the UK. So far 400 CEFIC Tremcards have been published,[10] covering hazardous substances commonly transported in road tank vehicles. These are nominally available in nine languages, all standard phrases being translated into agreed forms.

20. In addition to CEFIC Tremcards, suppliers of non-scheduled chemicals frequently produce their own transport emergency cards as written instructions to drivers, with the primary object of providing advice to the emergency services

TRANSPORT EMERGENCY CARD (Road)

CEFIC TEC (R)-9a
May 1971 Rev. 1
Class V ADR
Marg. 2501 A, (a), 2° (a)

Cargo	**NITRIC ACID (above 70%) fuming** Colourless liquid giving off yellowish-brown vapour with perceptible odour Completely miscible with water
Nature of Hazard	The vapour poisons by inhalation Causes severe damage to eyes, skin and air passages Corrosive Attacks clothing May react with combustible substances creating fire or explosion hazard and formation of toxic fumes: nitric oxide
Protective Devices	Suitable respiratory protective device Goggles giving complete protection to eyes Plastic gloves, boots, suit and hood giving complete protection to head, face and neck Eyewash bottle with clean water

EMERGENCY ACTION — Notify police and fire brigade immediately

- Stop the engine
- Mark roads and warn other road users
- Keep public away from danger area
- Keep upwind
- Put on protective clothing

Spillage
- Contain leaking liquid with sand or earth, consult an expert
- Do not absorb in sawdust or other combustible materials
- If substance has entered a water course or sewer or contaminated soil or vegetation, advise police
- Use waterspray to "knock down" vapour

Fire
- Keep containers cool by spraying with water if exposed to fire

First aid
- If the substance has got into the eyes, immediately wash out with plenty of water for several minutes
- Remove contaminated clothing immediately and wash affected skin with plenty of water
- Due to delayed effect of poisoning, persons who have inhaled the fumes must lie down and keep quite still. Patient should be kept under medical treatment for at least 48 hours
- Seek medical treatment when anyone has symptoms apparently due to inhalation or contact with skin or eyes
- Even if there are no symptoms send to a doctor and show him this card
- Keep patient warm
- Do not apply artificial respiration if patient is breathing

Additional information provided by manufacturer or sender

TELEPHONE

Prepared by CEFIC (CONSEIL EUROPEEN DES FEDERATIONS DE L'INDUSTRIE CHIMIQUE, EUROPEAN COUNCIL OF CHEMICAL MANUFACTURERS' FEDERATIONS) Zürich, from the best knowledge available; no responsibility is accepted that the information is sufficient or correct in all cases
Obtainable from NORPRINT LIMITED, BOSTON, LINCOLNSHIRE
Acknowledgment is made to V.N.C.I. and E.V.O. of the Netherlands for their help in the preparation of this card

Applies only during road transport English

Fig. 6. CEFIC Tremcard

in the event of a transportation incident. A series of 'waste' Tremcards, drawn up by the Harwell Chemical Emergency Centre, are used nationally by the waste disposal industry for the same purpose. Tremcards produced in this way may subsequently be accepted by CEFIC or alternatively be replaced by CEFIC Tremcards when these become available.

MCA Chemcards

21. The Manufacturing Chemists Association in the United States was the first organization to launch a voluntary scheme, whereby hazardous chemical loads would be accompanied during transport by written emergency instructions. These instructions, each relating to a single substance, are called Chemcards (Fig. 7). Each card gives information on the name of the product and potential hazards, with advice for dealing with fire or spillage situations. Chemcards are also published as a manual[11] containing 86 entries. A companion manual of cargo information cards[12] is also produced by MCA to meet the requirements of the US Coast Guard for any regulated cargo traffic conveyed by barge.

Sources of specialist chemical information

22. In emergency situations it is essential that detailed specialist advice be sought swiftly, and wherever possible from the manufacturer of the product concerned. This is the principle of the UK Chemical Industry Scheme for Assistance in Freight Emergencies (Chemsafe) introduced by CIA in January 1974.[13]

23. One of the aims of Chemsafe is to develop a technical chemical hazard information and advisory service, available at all times to the public emergency services to provide assistance in the event of incidents where the product manufacturer is not known or cannot be contacted. This 'service' is provided in collaboration with the National Chemical Emergency Centre at Harwell, which was established by the Home Office and Department of the Environment to give centralized advice. A special emergency telephone at the centre is manned continuously and a team of technically qualified staff is always available.

24. The Harwell centre currently handles about five calls per week, most of which are due to chemical products inadequately labelled—frequently with just a trade name. In such situations the technical knowledge of the duty officers is, in itself, insufficient without reference to appropriate data. Whilst some information on the hazards of trade name products is available in reference books, for the majority of these materials no such data are published. The only satisfactory method of obtaining information for recognizing the many thousands of chemicals being transported, and their properties, is from the source—the manufacturer.

MCA CHEM-CARD — Transportation Emergency Guide | CC-56 February 1972

ISOPROPYLAMINE

Clear, colorless liquid; sharp, irritating odor. Gas heavier than air.

HAZARDS

FIRE — Extremely flammable. May be ignited by heat, sparks or open flame.

EXPOSURE — Vapor extremely irritating. Liquid causes severe burns.

IN CASE OF ACCIDENT

IF THIS HAPPENS ⇩   For assistance, phone **CHEMTREC** toll free, day or night **800-424-9300**   **DO THIS** ⇩

SPILL or LEAK — Shut off ignition. No smoking or flares. Keep people away. Keep upwind. Shut off leak if without risk. If necessary to enter spill area, wear self-contained breathing apparatus and full protective clothing including boots. Flush area with water spray. Run-off to sewer may create explosion hazard; notify authorities.

FIRE — Do not put out fire until leak has been shut off. On small fire use dry chemical or CO_2. On large fire use water spray. Wear self-contained breathing apparatus. Tanks in massive flame contact may rupture violently unless cooled by straight streams of water. Use unmanned monitor nozzle. Apply from side. Keep clear of tank ends. Otherwise, withdraw from area and let fire burn.

EXPOSURE — Remove to fresh air. If breathing is difficult, administer oxygen. In case of contact, immediately flush skin or eyes with plenty of water for at least 15 minutes; remove contaminated clothing and shoes at once. Call a physician. Keep patient at rest.

© Manufacturing Chemists' Association, Inc., 1825 Connecticut Avenue, N.W., Washington, D. C. 20009, 1972. While prepared from sources believed reliable, the Association makes no warranty that the information is, in all cases, correct or sufficient. Printed in USA

Fig. 7. MCA Chemcard

25. Accordingly CIA and Harwell compiled a questionnaire (Appendix 3) which all chemical companies are asked to complete for each chemical manufactured, marketed or imported. The information requested is based on satisfying the needs of the emergency services. The information provided on each questionnaire is stored in a computerized data bank which at present covers over 7000 materials.

26. A document produced from a questionnaire can be retrieved from any suitable computer terminal over telephone lines. Having contacted the computer

Table 2.

	PERMANATE
Trade name:	Permanate
Company name:	Parton Chemical Co. Ltd, Northern Road, Parton, Warwickshire
Packaging:	25 kg paper sacks
Code marks:	186KP
Composition:	Potassium permanganate, 100%
Form:	Solid, crystals, dark purple-metallic sheen
Hazards:	Powerful oxidizing material. Explosive in contact with sulphuric acid or hydrogen peroxide. Reacts violently with finely divided easily oxidizable substances. Spontaneously flammable on contact with glycerine and ethylene glycol. Increases flammability of combustible materials. A strong irritant due to oxidizing properties, use breathing apparatus.
Hazard class:	UN serial no. 1490 UN hazard class 5.1.0 Hazchem code 1P Kemler 50
Spillage:	Wear breathing apparatus at all times with full protective clothing. Flood with water. Beware hazard of contaminated clothing on drying out.
Fire:	Flood with water.
First Aid:	
Knowledge:	Mr H. Weston
Phone:	Parton 49255, available 24 hours
Routes:	Parton to North London and Wales principally
References:	Technical data sheet 1, 2(a) enclosed
Compiler:	Mr J. Jones, Parton 49312
Date:	September 1974

it is only necessary to enter any series of characteristics of the material, followed by '?'. Such characteristics could include a trade name or even part of a name if that is all which is available, a code number, and a description of the product or its container. The example of a fictitious product given in Table 2 illustrates the output format for the data currently being collected.

27. At present the computer is only accessed by Chemical Emergency Centre staff but its ease of access by non-scientific personnel suggests a more general application. Experimental access to the computer data bank from different parts of the UK has demonstrated the technical feasibility of an 'on-line' system and of its direct use by non-scientific fire service personnel.

28. The computerization of data banks holding environmental information, especially in the fields of water pollution and toxicology, is now widespread. This is no doubt assisting, and being assisted by, the growth of data exchange systems and information networks. It is not possible within the scope of this Paper to discuss many of these, though it is of interest to note that the Association of Special Libraries and Information Bureaux (ASLIB) has produced a directory[14] listing more than 30 environmental data banks in Europe alone.

29. One European data bank embracing both environmental and 'emergency' information is the Environmental Chemicals Data and Information Network. ECDIN is a research project conducted within the EEC's direct action programme at the Joint Research Centre in Italy in collaboration with various national research centres. It is concerned with the international pooling of substantive information on chemicals, in particular toxic materials occurring in the environment as a result of human activity. Information held includes physical and chemical properties, production and use data, handling and transport requirements, dispersion and transformation pathways in the environment, toxicity and regulations. The ECDIN project is now nearing the end of its research phase and could become a substantive data bank in the future.

30. In the USA, the Environmental Protection Agency has established the Oil and Hazardous Materials—Technical Assistance Data System (Ohmtads). This is a computerized, on-line information file, designated to facilitate rapid retrieval of information on more than 1000 materials, and resembles the Harwell system in many ways. It includes information pertinent to spill response efforts, with a wide variety of physical, chemical, biological, toxicological and commercial data. However, the greatest emphasis is placed on the deleterious effects these materials may have on water quality. Direct access to Ohmtads from Europe is now possible through Brussels, and the data base may also be obtained on microfiche for manual use.

31. Another computerized data retrieval system in the USA is operated by the US Coast Guard as part of the Chemical Hazards Response Information

System (CHRIS) organization.[15,16] CHRIS is concerned with protection of the marine environment and enables Coast Guard personnel and others to have access to essential information in emergencies involving releases of hazardous materials on water. It consists of four data manuals holding information on about 400 hazardous chemicals, a hazard assessment computer system and system support personnel.

32. However, the best known US source of information available in the event of a chemical transportation accident is the Chemical Transportation Emergency Centre in Washington. Chemtrec became operational in September 1971 and is operated by the MCA. It provides a 24 h service of immediate advice for those at the scene of an emergency, then contacts the shipper of the chemicals involved for more specific assistance and appropriate follow up. The Centre is operated by non-technical personnel who obtain information on an incident from an enquirer, then provide appropriate technical information from a manual data retrieval system prepared by technical experts. Fig. 8 shows an example of such a technical data sheet.

33. Chemtrec has two significant differences from the Harwell Chemical Emergency Centre.

(a) Chemtrec is ready to receive all emergency calls throughout the USA using a single 'toll-free' telephone number, whereas Harwell provides response in longstop situations only, using an ex-directory line available only to the emergency services;

(b) the communicators' at Chemtrec provide information from a hard copy document and do not enter into a technical dialogue; Harwell's duty officers are scientifically trained and are able to discuss problems with enquirers and provide assistance from their chemical training.

34. Many individual companies have well organized response capabilities for their own products, some of which preceded Chemtrec by several years. Chemtrec collaborates with them and enhances their effectiveness.

35. In Canada, the Canadian Chemical Producers' Association operates a Transportation Emergency Assistance Programme (TEAP) through regional teams prepared to provide telephone advice and field assistance.

36. Environment Canada has developed a comprehensive computer-based environmental emergency management support system. This includes a national emergency equipment locator system (NEELS) which identifies the location of the required equipment nearest to an incident. Another computer system, the Water Resources Document Reference Centre (Watdoc),[17] can be used to search applicable chemical, biological, engineering or water resources abstracts from about 200 000 citations. Chemical and hazard identification is obtained through

CUMBERLAND & FEATES

1.2 COMMON NAME: Methyl Alcohol * Carbinol *
SYNONYMS: Wood Alcohol * Anti-freeze *

1.3 PHYSICAL FORM & APPEARANCE: Colorless, water-white liquid

1.4 ODOR: Alcoholic to pungent oily odor

1.5 EFFECT WITH WATER: Mixes with
1.6 SHIPPING OR B/L DESCRIPTION: Methanol - Flammable Liquid

SHIPPER OR MANUFACTURER: SEE REVERSE

NATURE OF PRODUCT: Flammable liquid that has little or no immediate health hazards.

DOT CLASS	F.L.
THI GUIDE	70A

HAZARDS

FIRE: Flammable, may be ignited by heat, sparks or open flame. Vapor-air mixtures can be explosive.

EXPOSURE: Inhalation of vapor and repeated contact of the liquid or vapor with the skin is harmful and effects may be cumulative.

IN CASE OF ACCIDENT

SPILL OR LEAK: Shut of ignition. Keep people away. Keep upwind. No smoking or flares. Shut off leak if without risk. Wear self-contained breathing apparatus. Dike large spills and remove by pumping into a salvage tank. Use water spray to knock down vapor. If removal is impossible, or for small spills, flush area with water spray. Run-off to sewer may create explosion hazard; notify fire, health and pollution control agenicies.

SD-22
CC-69
CIC-84

FIRE: On small fire use dry chemical or carbon dioxide. On large fire use water spray or alcohol foam. Cool containers with water if exposed to fire.

EXPOSURE: Remove to fresh air. If not breathing, apply artificial respiration, oxygen. If breathing is difficult, administer oxygen. Call a physician. In case of contact with skin or eyes, remove contaminated clothing and cleanse skin with soap and water. Flush eyes liberally with water and obtain medical attention.

F.P. 61°F
I.T. 878°F
V.D. 1.1

DISPOSAL: Incinerate in properly designed burner; or submit to contract disposal service for reclaim or disposition.

*ALSO FILED UNDER THIS NAME   T-TRADE NAME

Fig. 8. Chemtrec data sheet

Ohmtads in the USA, the only 'non-Canadian' part of the computerized emergency system.

Summary and conclusions

37. Some of the systems used to identify chemicals and their hazards following a transport incident have been reviewed here. Many other systems exist, but tend to operate on a local basis and have not been considered.

38. Developments in data handling techniques during the past few years, together with a more conscious approach to the environmental significance of spills of hazardous material is undoubtedly leading to an acceptance of the need for speedy chemical and hazard identification in emergency situations.

39. One of the common difficulties shared by all information systems is in obtaining and disseminating data. This applies whether the information is used to provide a classification or for inclusion in a data retrieval system. The questionnaires included in the appendices illustrate the similarity in information being sought by different committees, centres or organizations. This is encouraging, since it shows there is some approach to a consensus view on information needs, but it can also involve duplication and inefficiency. The chemical industries are invariably the source of this information, and perhaps serious consideration should be given to achieving closer collaboration between all parties, national and international, to reduce the effort and cost of data preparation.

References

1. CETGD. *Transport of dangerous goods.* Recommendations prepared by the Committee of Experts on the Transport of Dangerous Goods. United Nations, New York, 1976, ST/SG/AC, 10/1.
2. IMCO. *International maritime dangerous goods code.* Intergovernmental Maritime Consultative Organization, London, 1975.
3. DEPT. OF TRANSPORT. *European agreement concerning the international carriage of dangerous goods by road (ADR).* HMSO, 1976.
4. DEPT. OF TRANSPORT. *International regulations concerning the carriage of dangerous goods by rail (RID).* HMSO, 1977.
5. IATA. *Restricted articles regulations.* International Air Transport Association, Geneva. 19th Edition, 1976.
6. HSE. *Hazardous substances (conveyance by road) tank labelling regulations* 1977 *and Transport hazard information rules.* Consultative Document. Health and Safety Executive, London, 1977.
7. CIA. *Hazard identification: a voluntary scheme for the marking of tank vehicles conveying dangerous substances by road and rail.* Chemical Industries Association, 1976.

8. CIA. *Hazchem codings, allocated by the Joint Committee on Fire Brigade Operations and confirmed by the Health and Safety Executive.* Chemical Industries Association, 1976.
9. CIA. *United nations substance identification numbers for the transport of dangerous goods.* Chemical Industries Association, 1976.
10. CIA. *Transport emergency cards* (4 vols.). Chemical Industries Association.
11. MCA. *Chem-card manual. A compilation of guides for the safe handling of chemicals involved in highway emergencies.* Manufacturing Chemists Association, Washington, USA, 1972.
12. MCA. *Cargo information card manual for bulk dangerous cargoes.* Manufacturing Chemists Association, Washington, USA, 1970.
13. CIA. *Chemsafe: a manual of the chemical industry scheme for assistance in freight emergencies.* Chemical Industries Association, 2nd Edition, 1976.
14. ASLIB. *Data bases in Europe. A directory to machine-readable data bases and data banks in Europe.* Association of Special Libraries and Information Bureaux London, European User Series 1. 2nd Edition, 1976.
15. ANNAN D. S. et al. *Preliminary system development chemical hazard response information system (CHRIS).* US Coast Guard (NTIS AD-757-472), May 1972.
16. BROWN G. H. Implementation of a chemical hazards response system *(CHRIS). Proc. Conf. Control of hazardous material spills,* Oil Spill Control Association/EPA. New Orleans, 1976, 103–4.
17. IWD. *Water Resources Document Reference Centre: Watdoc.* Inland Waters Directorate, Environment Canada, March 1975.

PAPER 2

Appendix 1. Information sheet for new substances to be added to the lists

Chemical name:

Other names:

Formula:

Proposed Classification: Class _____ Group _____

Physical state: Solid ☐   Liquid ☐   Gas ☐

                 °C   °F

   Boiling point

   Melting point

Fire hazard:

Flash point _____ °C _____ °F   Open cup ☐   Closed cup ☐

Toxicity:

   LD_{50} oral _____ mg/kg

   LD_{50} dermal _____ mg/kg

   LC_{50} inh. _____ ml/m^3 (ppm) _____ mg/l

Corrosion: Animal (skin) Positive ☐   Negative ☐

          Steel _____ mm/year, _____ in/year

          Aluminum _____ mm/year, _____ in/year

Oxidizer: Strong ☐   Medium ☐   Weak ☐

Self-reactive or polymerize: Violent ☐   Medium ☐   Low ☐

Other hazards

Remarks

CUMBERLAND & FEATES

Appendix 2. Data required for Hazchem coding   Draft
Name and/or trade name   ....

1. Physical data*
   (a) Physical form and temperature as carried (if solid describe)
       ....
   (b) Physical form at 20°C and 760 mmHg................
   (c) Melting point, °C................................
   (d) Boiling point, °C................................
   (e) Vapour pressure mmHg at 20°C.....................
   (f) Flash point °C ..... Open cup/closed cup (delete as necessary)
   (g) Density at 20°C ....
   (h) Miscibility in water w/w at 20°C ....
   (i) Vapour density at 20°C ....
   (j) Other relevant comments (or expansion of above)...............
       ....

2. Toxicology*
   (a) Is it significantly corrosive to the skin? ....
   (b) Is there a serious chronic toxicity effect by any absorptive route?......
   (c) Is it significantly toxic by skin absorption? (ignore simple defatting)....
   (d) Is it significantly toxic by inhalation? ....
   (e) What is its current TLV (as recommended by ACGIH) ppm? ....
   (f) Is there a risk of permanent damage to the eyes?................
   (g) Would there be a serious toxic risk even on dilution in water (> 10 : 1)?
       ....
   (h) Other relevant comments (or expansion of above)...............
       ....

3. Stability and reactivity*
   (a) Is it combustible? ....
   (b) Is it inherently unstable, e.g. if heated? ....
   (c) Is it a strongly oxidizing substance? ....
   (d) Does it react violently with water? ....
   (e) Other relevant comments (or expansion of above)...............
       ....

Classification   UN no ☐   UN class ☐ ☐   ADR class ☐ ☐

*In the case of mixtures, details of composition may help: ....
....

Please also attach a copy of your Tremcard or other instructions in writing provided for drivers. If none exists, please tick ☐

PAPER 2

Appendix 3. Chemical Industries Association Limited. Chemsafe— chemical product emergency information

1. Name of company...

   Address...

   ....

2. Product name
   (i.e. name given
   on label/package)   [   ]   Manuf'd. ☐☐☐
                                                         Marketed
   Alternative names used ....   ☐☐☐
                                                         Imported
   (if any)   ....   ☐☐☐
                                                         Please √ box

3. Code marks (if any)   [   ]

4. Approved chemical name of   %
   constituents.
   (with approx. concn. if mixture)

5. Physical form | Solid | | Colour: |
                 | Liquid | | Other features: |
                 | Gas | | |
                 Please √ box

6. Type of packaging   size and description
Sack	
Drum	
Bulk	
Other	

7. Hazards (<u>brief</u> description and handling precautions)

Flash point   [   ]
(if flammable)

8. Product/hazard classifications (if known)

UN serial no.
UN hazard classification (division and subdivision)
Kemler code
Hazchem code

9. Recommended emergency action in event of:
   (a) spillage

   (b) fire (e.g. extinguishing media)

10. First aid treatment

11. Name of individual organization with specialist knowledge
    ....
    Emergency telephone number [   ]
    Availability (days and hours) [   ]

12. Principal transport routes:

13. *Literature references (e.g. technical data sheets giving additional information)

14. Name and telephone number of compiler (in event of any queries)
    ....

   *Wherever possible such publications should be included with this completed form.

Discussion: Papers 1 and 2

Mr M. W. Pullin (Fire Section, Scientific Branch, GLC)

My section of the Scientific Branch of the Greater London Council gives scientific and technical advice on operational problems to the London Fire Brigade (LFB). Together we produced the original Hazchem scheme and have worked on its subsequent development with the various government departments since; and it is good to know that the UKHIS Hazard Information Panel which includes the Hazchem action code will soon be a statutory requirement for tank vehicles. These vehicles are, however, only part of the total load of hazardous goods transported by road and rail.

2. In 1977 attendances have been made by the LFB to 240 incidents where hazardous substances have been or were thought to have been involved, though many of these were of a minor nature. At 22 incidents of a total of 62 occurring on the public highway there was a need for the brigade to ask for scientific assistance to make the situation safe. Three of these involved tankers, and fortunately each was of a minor nature (petrol tanker incidents not included). One of the tankers was labelled with the UKHIS hazard identification plates on the sides of the vehicle but not on the rear, and the other two were not clearly labelled at all. Seventeen incidents involved lorries loaded with 5, 10 and 40 gallon containers of acids, alkalis, solvents, etc. The labels on these varied from good identification, through a bad label, just a trade name and number (often with no manufacturer's name), to no label at all. The other two incidents involved bags of powder, one of which resulted in a white powder being spilled from 50 kg bags labelled 'phosphorus pentoxide'. The powder was later identified as nothing more than polythene.

3. These statistics confirm the view that incidents involving tankers are few, and because of the many safety features incorporated in the design of the vehicle those incidents which do occur tend to be minor in nature. It is the lorry loaded with various containers and bags which seems to give the most trouble, although the area involved is usually close to the incident. The causes of the incidents attended in 1977 were mainly bad stowage, damaged containers being sent out from manufacturers or distributors, and caps not being tightly screwed down.

4. In view of the problems caused by these containers, it would be useful if we could learn what regulations are being prepared by the Health and Safety Executive (HSE) to cover the present inadequate labelling in this area of the transport of hazardous loads. Perhaps the answer is an extension of the use of

DISCUSSION: PAPERS 1 AND 2

the UKHIS plate, as this is already understood by the emergency services. I would prefer to see uniformity in labelling so that one label could meet the needs of road, rail and sea, and even possibly storage.

5. Since the firemen's strike began,* there has been only one very minor incident to which we have been called on the public highway. This is not the normal frequency, and I wonder if the Harwell centre has experienced different phenomena during this time. I am hoping that the strike has made all transporters load their lorries and tankers more carefully, and drivers drive more cautiously.

6. The Authors of Paper 2 describe the facilities available at Harwell, and assert that these can be easily used by non-scientific fire service personnel. Have any relevant tests been carried out? Some discussion has taken place on the setting up of a special fire service data retrieval system containing only information relevant to dealing with incidents. Is this suggestion practicable, since clearly fire services need different information from, say, a water authority. If Hazfile is to be developed further, would a governmental committee with all relevant bodies represented be the best way to do this, or should it be done by the Harwell staff?

7. The various labelling and data retrieval schemes now in use have been described in considerable detail in Paper 2, and the Authors have gained much practical experience in this field in recent years. With this experience, I wonder if they have any views on whether, say, the Hazfile scheme is the best method or whether if they started again they would use a new set of criteria to produce the ideal data retrieval system. For example, one aspect that seems to be missing in the 'Permanate' data sheet are the relevant methods of decontamination of protective clothing. All the references in the Paper are to labelling schemes in Western Europe. Are there any similar schemes in Eastern Europe?

8. For any data system to be of use, the correct information must obviously be fed in. This requires the manufacturers to provide information on the various properties of the substance. The data are collected on a questionnaire. Each authority requiring data has its own questionnaire, each differing slightly from the next. The result is that the manufacturer has to fill in several similar forms. Closer collaboration between all parties is an urgent requirement. However, will the problem of commercial confidentiality become more severe if a national or even international questionnaire is produced?

Mr D. Knight (Wessex Water Authority)
I represent one of the several authorities which have to deal with the product if it

---

*Eds note: The conference took place during the national firemen's strike of 14 Nov. 1977–16 Jan. 1978.

gets washed into the drain. Before raising one or two points about Paper 2 it would be as well to consider the routes followed by liquids that get washed from the road, and the hazards that they might present to other people.

10. When the liquid goes down the gully it may go either to a sewer or to a river. In a sewer there could be people working, especially if it is in a very large urbanized area. Certainly it will turn up at the sewage works. This is a biological treatment system designed to purify the sewage, and any chemical could affect these processes. Given sufficient warning at a large sewage works, it may be possible to divert the pollutant into storm holding tanks and deal with it at some later stage. Far more concern should be shown about any discharge getting into a river. The river is not just a body of water flowing from one point to another and thence out to sea; it is used by a wide variety of people. People could be swimming or boating on it, cattle may be drinking from it. The industrialist may be abstracting water for cooling, steam-raising, or for process purposes. More important from the water industry's point of view, water may be abstracted from that particular river. If there is sufficient warning the intake can be shut off until such time as the particular contamination has passed the abstraction point.

11. I do not believe that the river or the sewer should be protected under all circumstances. There are cases where the protection of life and property takes precedence over the protection of water. However, the failing of the Hazcode is in its simplicity. It does not distinguish between the spillage in an urban area where the protection of life and property may be necessary and the spillage of the same chemical in a rural area with little or no habitation but where there are abstractions. The Hazcode does not give advice to the firemen on whether they should concentrate on the protection of property or protection of the watercourse.

12. In Paper 2 §13 there is reference to a 'decision tree', which has been evolved over the past 2-3 years. Would the Authors comment on whether this is the reason why, in the recent Health and Safety Executive Consultative document on the Transport of Hazardous Goods, of the 44 chemicals listed as poisonous, 41 are 'contain' and, three (diquat, acrylamide and dimethyl formamide) are 'dilute'. Were these codes produced at the early stages of the Hazchem Coding Committee?

13. I understand that the Hazchem Code itself is primarily designed for the firemen's protection and to enable them to take first aid action within the first 15 min. or so of a particular spillage taking place. This is part of an augmentation scheme whereby additional information is available through the Tremcards or through the Chemsafe scheme. It is interesting to note that the nitric acid Tremcard shown in Fig. 6 of Paper 2 says 'contain'. The Hazchem Code for

DISCUSSION: PAPERS 1 AND 2

nitric acid is 'dilute'. The firemen could be in a quandary when, having acted on the Hazchem Code, they receive additional information from the Tremcard to 'contain'. Their initial action will, quite likely, have made the situation worse since not only will there be more liquid to contain but the firemen will have to obey apparently conflicting instructions.

Dr P. L. Rose (Severn Trent Water Authority)
I would like to endorse the previous speaker's comments. One or two occurrences have come to Severn Trent's notice where the recommendation of washing to the drain has considerable risks. While we accept that public safety must come first, there seems a little scope for moderation of this advice, depending on the circumstances and the location of the spillage.

15. I would like to examine a major shortcoming of the UKHIS system. There was recently an accident on the M5 just south of Worcester, when a 3000 gal. tanker overturned. The tank was ruptured and the whole of its contents went on to the hard shoulder and into the drainage system. One of the disadvantages from a water authority's point of view is that every mile of the motorway is well drained by pipes which are taken to the nearest watercourse and therefore anything which is spilt gets to that watercourse rather faster than we do. This particular spillage was identified from the UKHIS plate on the rear of the vehicle as trichloethylene, and this chemical was sought in the nearby stream which was a feeder to the River Severn, upstream of two major water supply intakes. Shortly afterwards further information was passed that it was in fact methylene chloride, and we took the appropriate steps.

16. It was afterwards found that that particular vehicle had three carriers, one on each side and one on the rear of the vehicle, and a total of eight double-sided UKHIS plates, plus three paper labels, covering at least five different chemicals. The Tremcards carried in the cab covered three of these and one other, so that there was a total of six different chemicals covered by the plates and the Tremcards. It was fairly certain that the vehicle as it was operating down the motorway showed the right designation on all three of the external plates, but of course when it turned over most of them fell out. The only one that remained (because it was bent enough to jam into the system) showed trichloethylene.

17. The chemicals are all, in fact, chlorinated hydrocarbons all with the same Hazchem code, so that as far as the emergency services were concerned there was no incorrect information given, but there were quite considerable differences so far as other hazards were concerned. The boiling points of the six possible chemicals, ranging from 42° to 121°C presented very considerable differences in volatility. There is also a considerable difference in their solubility in water,

although all are fairly insoluble by chemical standards, the highest solubility being 2%. Such contaminants in water are significant at the parts per million level and the solubilities range from 10 ppm to 20 000 ppm.

18. The hazard of the identification error in this incident was that trichloethylene is one of the least soluble and would sit on the bottom of a small stream being slowly washed away into the water. Since there would be a considerable dilution before it got down to the water supply intakes it is probable that no emergency action would be needed other than to suck out large accumulations from the bottom of the stream. Methylene chloride has the highest solubility of the six chemicals, about 20 000 ppm, and much more urgent action was needed. In fact the stream was dammed and the contaminated water pumped away for disposal. The toxicity range is also rather different—a factor of about 1000 between the most and least toxic to small mammals.

19. I would like to ask those who may be involved in improving the UKHIS system if the regulations can include some coverage for the problem of multiple plates. I understand from the operator of the tanker fleet that this case involved a tanker on a round trip, loading and delivering several different products, and the easiest thing is for the operator to carry all the necessary cards and arrange them so that the correct one is on the top. But this does not cover the possibility of an accident in which the vehicle overturns. Equally, the Tremcard system does not cover the possibility which it was meant to cover, where the driver of the vehicle is incapacitated.

Mr D. T. Corbett (Technical Director, Fospur Ltd)
The point has emerged that the identification of a potential hazard is extremely important as the first step in coping with a dangerous situation. In fact, this information may in certain circumstances take hours when minutes are really important.

21. Some time ago my company was approached by Martin Pitt of the University of Aston and Robert Keen of Bristol Polytechnic who had developed a mini-laboratory with the idea of forecasting very quickly the categorization into which a hazard could be placed so that the acting emergency officer or fire brigade had some immediate information. It included dangerous fumes, corrosive substances, behaviour of the spillage with water, flammable or explosive properties and potentially poisonous characteristics.

22. The object of the unit was to have something portable and easily used with simple instructions, so that a non-technical man could follow the scheme of operation. The interpretations to be drawn from the tests were also carefully laid out, so that if a specialist was not able to be on the site immediately, he would have by telephone results of a series of chemical tests, from which he

DISCUSSION: PAPERS 1 AND 2

could make assumptions possible from a description of the appearance of the substance.

23. As this unit was developed, much work was done in connection with the West Midlands Fire Service who gave a lot of advice on the styling and format of the instruction manual. It would have been useful, of course, to have linked the conclusions of the tests into the Hazchem coding, but that was not possible. The ultimate aim was to come forward with a general categorization of a hazard.

24. Emergency services and industry need something which gives a very quick answer. It is clear that there is a gap between when the accident occurs and when one can actually diagnose the problem.

25. I would like to have the Authors' comments on how such a system can fit into the overall hazard control system and the hazard identification process which is currently being used.

Dipl.-Chem. W. Hopfner (Bundesanstalt für Materialprüfung, Berlin)

In Germany the Bundesanstalt für Materialprüfung (BaM) is the officially authorized licensing office for tank containers used in the transport of hazardous materials. In order to assess the compatibility between transported substances and container material we use the DECHEMA* table and the NACE† table, but there is a considerable amount of data that cannot be found in these tables, in particular that relating to corrosion, pitting, stress corrosion, cracking, etc. Welding, which according to the ADR regulations should be completely safe, is not investigated in many cases.

27. The BaM therefore circulated a number of governments, institutions, departments, organizations, etc. in Europe and America, asking what kind of data they used in connection with licensing tank containers for transporting dangerous substances. The answers were frightening: 'We have no experience; we have the same problems as you have. If you know anything, please tell us.' 'We do not look for corrosion; it is the responsibility of the driver.' 'We have no concrete data.' Customers who have wanted a licence have told us that in England and in Belgium it is very easy to get a licence, and that there is no limit on the impurities which may be transported, particularly water and chlorine ions. I feel that the licensing system should be more uniform.

Mr P. N. Anderson (Imperial Chemical Industries Ltd)

In the chemical industry we have always striven to be ahead of legislation, to embark upon codes of practice and set up good approaches to safety in advance of the legislators. It is our intention to stay ahead of the legislation and to this

---

* Deutsche Gesellschaft für Chemisches Apparate.
† National Association of Corrosion Engineers.

end we have instigated the Chemsafe emergency response systems throughout the industry; we have spread the use of Tremcards across Europe; we have given every assistance to local authorities where routing is promulgated, and we have embarked upon an extensive system of training for drivers involved in the movement of hazardous goods.

29. We have also attempted to keep designs of road tanker and packages ahead of legislation, although in the case of the petroleum spirit tanker the legislation does not express a minimum but an absolute design, and therefore we are compelled to move such materials in less than satisfactory tankers by present day standards. There is a need for speedy reform of that piece of legislation.

30. Several speakers have referred to the information which is available to the emergency services in the event of an accident. This comes in several layers. First is the instantaneous information for which the fire services developed a coding scheme which has become known as Hazchem. This gives a few clear instructions to the fireman. It does not tell him what the hazard is but what to do. The next is the UN diamond label which indicates the prime hazard more quickly than a chemical kit and with less chance of going wrong. The third is the UN number of the name of the product. That gives the fireman a link to sources of expertise. Finally, there is a telephone number and an address which will give him what in my view is essential in a major accident. There is no substitute for an expert's advice. Every accident is different and to apply knowledge of the physical characteristics and hazards to each particular incident requires someone who is expert in the field.

31. The EEC organization is not the same as the UN organization in matters of the transport of dangerous goods. The EEC has a directive to do with the labelling of packages. The purpose, however, is not to deal with hazardous goods in transit or in an emergency but rather to instruct the user on the hazards of the materials. This means two different approaches to labelling: one for handling in an emergency and the other for handling under normal circumstances. These can be combined, but I would like to be assured by Mr Dunster that the EEC is not itself contemplating any directives relating to the transport of dangerous goods since this is well covered by the UN European organization.

32. I agree with the general conclusion of Paper 2. There are many sources of information, and co-operation between these sources has now become important. There is no doubt that there are subtle differences in the advice from different sources. That is not because people are ignorant: it is because they are striving to do the best they can to advise, but their concepts of the accident are subtly different. The Hazchem Coding Committee for example, has many times been faced with the difficulty of deciding exactly what sort of incident must be considered for coding of the chemical. It is impossible to code for all

eventualities, and the tendency of the Committee is to settle for an incident somewhere on the worst side of average.

33. There must be some co-ordination between the sources of information, and these sources must not proliferate, or the people who are demanding the information will become confused. And there is no substitute for the expert on the spot.

Mr H. D. Peake (Director of Technical Services, London Borough of Ealing)
Local authorities are at the receiving end to a very great extent. Ealing has just spent £100 000 clearing up after an incident involving a petroleum tanker which had fallen over as it went round a roundabout. There are baffles inside tankers, but even so, 5000 gal. of petrol were discharged into the surface water sewerage system. Four hours later and 2 miles away there was a tremendous explosion in two 6 ft dia. sewer pipes. Houses had to be evacuated; schools had to be tested because petroleum fumes had entered them; social services had to be called out to deal with the evacuees, and the repair bill was £100 000.

35. What have we learned from this incident? First, there must be provision in all legislation for local authorities to be included in the liaison team with emergency services. It is essential that they are informed of what has happened, the nature of the chemical, for example, petrol, hydrochloric acid or ethanol, all of which have featured in incidents in Greater London. One of the problems is to notify men working in the sewers, who may not be aware that there has been an incident 2 or 3 miles away which may endanger their lives. Social services must be informed by the local authorities that they may have to take action in regard to rehousing people. Immediately there is an incident, the main drainage engineer must be at the scene so that he can advise on sewer runs, so that action can be taken at appropriate points to deal with the chemicals involved.

36. The next and most important point is illustrated by something which arose in a local authority area where I served. The incident involved a discharge of chemicals into the drainage system. This resulted in a chemical reaction with trade effluents from an industrial estate, producing a complete blockage from deposited carbonates and sulphates. Nobody, of course, would accept responsibility, and again the local authority had to find the money to provide for relaying certain sections of the sewers.

37. The local authority must be provided with up-to-the minute information, accessible at all times of the day, concerning movements of hazardous loads, chemical content, method of treatment if spilled, protective clothing required, and any additional measures.

DISCUSSION: PAPERS 1 AND 2

Mr K. N. Palmer (Fire Research Station, Borehamwood)
With data systems such as Hazchem one can only get out of the system as much information as was originally put in. One of the problems faced at the Fire Research Station (FRS) is the generation of information to feed into this sort of system. The problem arises with chemicals which have been well known in the laboratory for generations, but are suddenly required in an industrial process, and have to be transported by the tank load instead of by the Winchester bottle. The question then arises: if there is a spillage, what sort of fire extinguishing agent should be used with this material? Quite often the scientific literature provides no answer. I should like to quote one example.

39. Someone came to the FRS with a chemical well-known to A level students which contained chlorine, and asked what would happen if a tanker had a major spillage with fire following. After examination of the structure of the chemical and having found very little on its fire properties in the literature, it was conceivable that it would generate phosgene either during burning or immediately after the flame had been extinguished. Controlled small-scale experiments demonstrated that phosgene was not in fact produced, but it did produce vinyl chloride monomer, which has been identified as a potential carcinogen.

40. I would therefore stress that the means by which the information for these systems is gained must be strengthened, and that it is perhaps preferable to have some sort of systematic and safe method of progressing so that information concerned with new chemicals is generated internationally. A specification of fire extinguishing agents for new chemicals is to some extent done at the moment on guesswork. If someone wants to check the guesswork, he may run into trouble.

Dipl.-Chem. W. Hopfner (Bundesanstalt für Materialprufung, Berlin)
I should like to ask the Authors of Paper 2 if they are co-operating with DABAWAS*. DABAWAS is a German data bank set up to deal with accidents involving dangerous substances and contains 5000 of these substances.

Mr S.A. Berridge (Berridge Incinerators Ltd)
I am particularly concerned with the incineration of toxic and hazardous wastes.

43. I would like guidance from Mr Dunster on the transportation of such wastes. Chemical products are normally made to very strict specifications and under strong quality control. Toxic and hazardous wastes are under effective control at the disposal end because they are dealt with either by chemical means or by incineration, and the people concerned want to be alive the next day. The Alkali Inspectorate, local health authorities and other authorities are also monitoring matters to make sure that the necessary rules and regulations are observed.

---
* Datenbank wassergefährdender stofte.

DISCUSSION: PAPERS 1 AND 2

44. In my experience there is virtually no control as scientists and engineers understand this over the production and shipment of toxic and hazardous wastes. There is a removal notice and a statutory requirement for people to do what is known as the right thing, but this does not match up to reality. I received a removal notice for thin solvents, but when the 64 drums arrived on site I detected a distinct odour of mothballs, and the contents turned out to be naphthalene. On another occasion paint thinners were specified, but what arrived was a semi-solid matrix of milk bottles, boiler suits and tin hats with paint residues, which indicated no control whatever over the toxic and hazardous material which was supposed to be there.

45. A further incident involved waste methanol containing 3% mercaptans. These are detectable by nose at 1 part in 100 million, and there was a smell immediately the bungs were opened. Mercaptans are used for injection into natural gas which has no odour. The release of mercaptans brought a fleet of East Midlands Gas Board vans round the local mining village to find out where the gas leak was. An undertaking was given that no more of that particular material would go to that site, but a short time later another consignment arrived, supposed to have no odour, but in fact producing a pronounced odour of mercaptans. This proved to be caused by the fact that the tanker had been carrying material containing mercaptans and had not been cleaned out.

46. It appears that in this area quality control is virtually non-existent at the moment and presents a very great problem. The appeal I would make is for adequate control of the production and transport of toxic and hazardous wastes.

Mr F. J. Simmons (General Manager, Schloetter Co. Ltd., Pershore)
My firm is a manufacturer of chemicals which are transported in packs of 25–200 kg. One of our problems is that of mixed loads, which can be from 1–10 tons. From time to time we have to decide which diamonds to use which can be category 3–5, 6 or 8. The loads may contain up to 25% of any one of each of these four.

48. Another problem relates to the type of package and how much can be shipped in the size of package available. This presents a hazard when distributing material to customers. The raw material comes from manufacturers such as ICI, and we then add organic substances which are used in an electroplating process. How do we, as manufacturer of these organic additives, determine what to do if there is an accident involving one of our vehicles, especially when what we make may contain four, six or eight complex additives?

49. Where do we go in respect of the Health and Safety Act to protect not only ourselves as a manufacturer but the man in the street, the customer, and the people who are called on for appropriate action if drums come off a vehicle?

Such occurrences present problems which cannot be dealt with, because information is not in the data bank.

50. Under the Health and Safety Act, manufacturers have to label each drum indicating the category that the material comes into, using the diamond-shaped label, and also the protection which is applicable if there is spillage. Can manufacturers utilize this system to protect the man who comes on site to deal with an accident which may have occurred?

Mr S. C. Baker (Home Office Fire Service Inspectorate)
Many people appear to believe the Hazchem Committee's guidelines are arbitrary. The difficulty is that some other body has already suggested certain things that the emergency services should do. But if the Tremcards are studied carefully, it is clear that they are to deal with a small incident, because they give such a wide variety of options as to what to use in fire-fighting. These methods are not necessarily in strict priority order.

52. For instance, with certain chemicals, say a petroleum product, they suggest 'foam', and then give options of a dry chemical extinguisher, a halone extinguisher or water spray. The difficulty is that someone reading those four options will perhaps use the water spray because it is available, which would be the worst possible thing to do. Although the options are printed in order of preference there is nothing on the Tremcard to say that it is an order of preference. The Hazchem system tries to give the fireman an instruction which at least will not worsen the situation. For instance, a '3' indicates that he should use foam on a particular substance, but if he has not got foam he must go to a higher number and not a lower one.

Dr D. Train (Cremer and Warner)
The discussion so far has concentrated on ensuring that the proper executive and emergency actions are coded at least and possibly have regulations attached.

54. The professional bodies are ready to help in any further work which needs to be done. Whilst it is clear that some things should be done by Harwell or by the HSE, the professional bodies would welcome direction on how working groups may help.

Mr Dunster (Paper 1)
Regulations are being prepared for the compulsory marking of road tankers. The extension of this system to other loads of transport and to storage will be considered, but is far from simple. Even the use of the coded markings for road tankers has problems, and the action codes will, I suspect, never be able to distinguish satisfactorily between immediate action to avoid acute hazards and

DISCUSSION: PAPERS 1 AND 2

longer-term precautions, such as those to protect water courses. These longer-term operations must depend on a knowledge of the material and of the specific location of the accident.

56. The responsibility for seeing that the correct plate is shown, and remains showing after an accident, must rest with the carrier, suitably advised by the consignor. I am not impressed by the idea of a mini-laboratory. I doubt its scope, speed and reliability.

57. The international scene is firmly set by the United Nations and the specialised agreements for different loads of transport. It is always risky to make predictions about the European Communities but no directive on the transport of dangerous goods is in hand at present. They are concentrating on testing, packaging and labelling.

58. The transport of wastes will always be a problem and I doubt if there is a regulatory solution for the miscellaneous mixtures of toxic materials. There is a clear responsibility on the consignor to warn and advise both the carrier and the recipient about the hazards and precautions, and any system must depend on a close collaboration between these three bodies. Much the same is true of the transport of mixed loads, other than in the fairly simple case of multi-compartment tankers. It is wrong to expect that these problems will be solved by regulation alone, although the increasing requirements for labelling will certainly help. As for wastes, the consignor must keep in close touch with his carrier and make use of specialist carriers in some cases.

Dr Feates and Mr Cumberland (Paper 2)

Mr Pullin asked whether Hazfile is the best information system which can be developed. This is certainly not true. No system is perfect. Hazfile was developed about 3 years ago to meet an anticipated need. It filled a gap and has been seen to fill a gap. We would not wish to see it changed radically because there is always a tendency to modify systems. With 8000 compounds in the data bank and an eventual 25-30 000, the task of making a fairly minor change to the data base can be quite astonishingly large. As has been mentioned, a chemical engineering plant is inevitably obsolete by the time it is complete. Any data system, by the time it works, is also obsolete, but it works.

60. The question of who should develop a system does not matter. Harwell has tried to demonstrate one way but there are some basic requirements which need to be satisfied by any organization which does try to develop a system. It must be sensitive to the needs of the emergency services; it must have feed-back from the emergency services to ensure it is giving the right information in the right form; it must have access to the manufacturing industry who can up-date the information and check it—a very important component; it must understand

DISCUSSION: PAPERS 1 AND 2

the information system which is being used, be it computers or a manual system, and it should not be an organization remote from the development of that system. In the UK the Fire Research Establishment at Borehamwood might do this job very well, perhaps better than anyone else, if they wished to do it. Harwell can do it reasonably well but not perfectly: so could the London Fire Brigade. The problem is always one of balance.

61. Mr Pullin referred to the inadequate or improper labelling of containers and mentioned a P_2O_5 label on a bag which in fact contained polythene. This is something of which we have first-hand knowledge at Harwell. It is a difficult problem and one the Chemical Industries Association is most concerned about and is trying to eradicate, but once the package is sold the manufacturer has no more control over what it is re-used for.

62. He also referred to a drop in the number of incidents during the fireman's strike. This is also our experience. The number of calls to Harwell in the 4 weeks prior to the strike was 22, of which 13 were from the fire service and 9 involved road transport. From the 20 November to 15 December we have had 3 calls from the fire service from a total of 7, only one of which involved road transport. This is probably due in part to organizations taking even greater care to see that drums do not fall from the backs of lorries, as individuals are taking greater care at home to switch off all the appliances and keep the doors shut before going to bed.

63. On the contradictory instructions for fuming nitric acid, the recommendations in the Transport Emergency Code say 'contain' but also 'suppress the vapour with water spray'. In the early days instruction for nitric acid was 'contain', but following an incident illustrating that it would be much safer to get plenty of water on it and dilute it rapidly to reduce the amount of vapour this was subsequently changed. This shows that the Hazchem codes are not just produced philosophically. There is a practical element behind them. If following an incident a need emerges for the code to be changed, then the Coding Committee will look very carefully at that particular code. They have in some cases changed codes, as in the case of nitric acid, and in other cases they have decided that the initial code was in fact the best one.

64. With regard to the comments of Mr Corbett concerning Hazkit, the possibility of chemical 'black boxes' being used by operators without technical knowledge is worrying. The analysis of complex chemicals is not something which can be done adequately by such systems. We heard today of the problems that industry and local authorities have met in differentiating between a series of chlorinated hydrocarbons including methylene chloride and trichloethylene. It is doubtful if a kit could have coped with that problem. Could a kit cope with the problems of very low concentrations of mercaptans or the naphthalene also mentioned? It might tell if a spill is acid, alkaline or a cyanide, but usually the

identification problem is much more complex. Kits like the one described could justify a name like Hazkit—it could turn out a true 'haz-kit'.

65. The question of corrosion data is very relevant to a meeting such as this. If adequate corrosion data are not available for tank vehicles or containers the professional institutions should be looking into the matter. It should be available. Work on pesticide packaging suggests that there has been much research on the compatibility of pesticides with tin plate or the coating materials often used on tin plate.

66. We have become very involved at Harwell with the question of toxic and hazardous wastes. They provide a particular problem because of the diversity of materials, as Mr Berridge knows, and there have been discussions between Harwell and the Department of the Environment. A series of general Tremcards has been developed which goes some way to meeting these needs, and it might be possible to produce a complementary series of Hazchem codes which could be displayed on vehicles. But the nature of these mixed loads makes it much more important for the producer of the waste to ensure that he does know what he is transporting and that he labels his vehicles adequately in terms of the risk associated with an incident. Hopefully the Control of Pollution Act will go some way towards ensuring that this happens.

67. So far as international transport is concerned, the German Democratic Republic, Poland and Yugoslavia in Eastern Europe, being signatories to ADR, presumably are conforming to those labelling requirements for the scheduled chemicals, but the basic problem, as Harwell sees it is not so much the large consignment—the tanker or tanker container—but the small package or canister that is just marked with a foreign name. We at Harwell have spent an hour or two trying to identify a foreign name on a package only to find it was the name of the Italian shipping company. It is very difficult to identify materials without getting the information from the manufacturer. This is the prime job of the Chemical Emergency Centre in building up the data bank.

68. The question of decontamination of protective clothing is very valid. Almost certainly it was not included because no one appreciated its significance at the time. It is something which might be built in at a later stage although it would involve considerable effort in amending the existing data. The method for dealing with a spillage of Permanate might give a guide to the method for decontaminating clothing. This is a particularly difficult case because the material is highly oxidizing and the clothes might burst into flame.

69. Information at the site of an incident is very important. Following pending legislation it is to be hoped that the information needed to provide immediate response to an incident will be available to the emergency services in the majority of cases. Obviously there must be cases where labels are destroyed or improperly

fixed or are not there at all because the shipper has just not bothered. Local specialist advice should always be available to the emergency services. The suggestion that it sometimes takes 3 hours to get specialist advice is disturbing. We should ensure that that time is much reduced.

3. Design features for road vehicles

C. D. CERNES, MIMechE., Chief Engineer, Scammell Motors, Leyland Truck and Bus

The volume of goods carried on the road is increasing and proportionately higher quantities of hazardous and abnormally heavy or large materials are being moved. Reliability, durability, maintainability and product support have been significantly improved and provide better availability and confidence, and consequently better profitability for the operator and customer. Compatibility with other road users requires attention to road speed, acceleration, retardation, and stability in order that the driver is able to utilize better the primary safety of the vehicle. Load security is dependent on safe loading practices, knowledge of the vehicle capabilities and load to body and body to chassis security. Specific hazardous goods such as spirit, or abnormally heavy or large loads require special design requirements.

The improvements in communication and of power generation in general enabled manufacturers to be freed from their dependence upon rivers and communications by sea. The development of railways meant that manufacturing units could be built in areas most suited for population expansion, and improved communications allowed access to raw materials and rapid distribution of finished products. Industrial development has gone hand-in-hand with development in communications, and in spite of recent problems with the environment and supplies of fossil fuels, it is generally considered that this is still an expansionist era.

2. It is, therefore, important that the UK continues to develop road transport, and consequently trucks, by taking the best features from all parts of the European market and concentrating efforts on those aspects which improve the efficiency of the vehicle in performing its task, whether that be the transport of people, general industrial cargoes or hazardous materials.

Legislation

3. The public demands, and the manufacturers and operators accept their social responsibilities. These are largely embodied in legislation, called the Construction and Use Regulations, split into the two sections dealing with the construction of vehicles by the manufacturer and the use of these vehicles by opera-

tors. These legislative constraints are gradually being replaced by a common European standard. Where a European standard covers an aspect of legislation embodied in the national legislation, then the national legislation will ultimately be replaced by the European standard according to an agreed programme.

4. European legislation, that is EEC legislation from Brussels, is based on the United Nations legislation—ECE, from Geneva. The framing of legislation was covered quite well during a symposium at the 1974 Motor Show.[1] The most useful aspect of legislation is its definition of interface problems between the truck and its load, and the laden truck and the road, and also its restriction on operations that compromise safety for the sake of increased profit.

Vehicle requirements

5. Since the vehicle specification is at the heart of the vehicle design it is necessary to consider not only the performance, the unladen weight, the gross vehicle weight and the body dimensions when forming a conclusion on its suitability for the carriage of hazardous materials, but also to consider reliability, durability, maintainability, the availability of parts support, ergonomics of the driver environment, the availability of auxiliary power sources, manoeuvrability, compatibility with other road users, load security and the security of the body to the chassis.

6. The design parameters for a truck require a careful examination of the variety of tasks for which any one model can be used. These include flat-bodied general haulage machines, truck mixers, tankers or three-way tippers up to 32 ton GVW, articulated tractors up to 30 ton 8 × 4 rigid chassis or truck-based passenger chassis. This selection of the more common applications illustrates the ubiquity required from a basic model range to accommodate long length and asymmetric loading of platform models, roll stability requirements for mixer chassis with high centre of gravity and offset load, ride requirements for passenger chassis, frame strength and stability on tippers, and packaging requirements of the articulated vehicles. Manufacturers have a unity of interest with the vehicle operators in reducing the part number count and consequently the spare parts count, and increasing part volumes so that more sophisticated tooling can be employed, thus removing quantity constraints and enabling improvements in parts availability and repeatability and quality.

7. The whole life cost of the trucks is, with the trend towards premium vehicles, of much greater importance than the purchase cost. Down-time, especially unscheduled down-time, is very expensive and can produce problems when transporting volatile materials.

Reliability

8. Reliability is a very difficult target to meet involving many different requirements, including:
   (a) the initial design, built on a comprehensive knowledge of the design parameters backed up by experience of the unit in-service performance;
   (b) adequate testing to ensure that there are no performance problems;
   (c) tooling investment, appropriate to the requirement of the part, to obtain repeatability;
   (d) careful market introduction, involving user trials and monitoring of early production samples, to ensure compatibility with development results;
   (e) early service feedback of significant failure patterns;
   (f) quick solutions to identified problems;
   (g) early introduction of production solutions to early problems.

9. Reliability is only obtained by meticulous attention to detail in every phase of design, the will to succeed, and a great deal of experience. A truck is immobilized if a headlamp switch fails just as effectively as if an engine fails.

10. Since the modern truck represents a considerable investment on the part of the operator it is necessary that it maintains its reliability over a long operating life, and durable vehicles contribute as much to improved whole life costs as reliability. A truck which operates reliably for 150 000 miles but is then beyond economic repair has a poor re-sale value and represents a poor investment to the operator.

11. In order to reduce off road times and to increase the availability of the more expensive modern truck it has been necessary that even more attention be paid to maintainability. This aspect is controlled both during the design of the truck and during its pre-launch development phase.

Product support

12. Parts support, like maintainability, is aimed at keeping trucks on the road and coping with the consequences of an unscheduled failure in addition to scheduled parts replacements. There are some requirements for the carriage of hazardous materials where the operation requires specific design features. It is then necessary for the manufacturer to make special provision for parts availability and service engineers and for the training of drivers and maintenance staffs. In these instances, it is not always possible to stock spare parts or have available specially trained staff at all the distribution and service points throughout the UK or abroad. It is necessary then to provide a 'vehicle off the road' (VOR) facility. This VOR service is recognized throughout the manufacturers'

organizations, and speeds the availability of parts or service by special delivery vans or air freight as necessary.

Truck design

Horsepower

13. The power requirements of trucks have become more difficult to define recently since improvements in fuel consumption, reliability, and operating speeds can be achieved with power in excess of the legislative minimum. European legislation varies from 6 brake hp/ton to 10 brake hp/ton depending upon the domestic terrain. This, however, makes it difficult to provide rationalized ranges of trucks and encompass the economies of scale associated with a European market. Some of the legislation encourages more power but, since engines are efficient over a limited range, the torque back-up may suffer. In many instances a well-matched lower power unit and transmission can be a more efficient transporter of materials than a badly matched high power unit.

14. Currently power unit design is regulated by the power to weight regulations of the national authorities and by current and proposed European regulations governing vehicle noise and emissions.

Weight

15. The unladen weight of a chassis remains an important design parameter, since any increase in unladen weight directly reduces the payload capability. Many operators, however, are prepared to accept some increase in unladen weight in the interests of improved reliability and consequent vehicle availability.

Capacity

16. The body dimensions are normally constrained by the vehicle configuration chosen. Rigid vehicles are unlikely to be able to provide a free body length of more than 27 ft within a manoeuverable configuration, and if the density of the material to be handled dictates, for economic or some other constraint, a larger overall length, it is necessary to utilize articulated vehicles of either semi-articulated, drawbar, or double bottom configuration. The latter are not legal in the UK.

Stability

17. Stability is of importance whilst accelerating and decelerating, laden and unladen, on up and down gradients, on adverse cambers and on poor surfaces. The problems associated with rigid, articulated, drawbar and double bottom outfits are all different. The extreme situation of a semi-articulated vehicle

CERNES

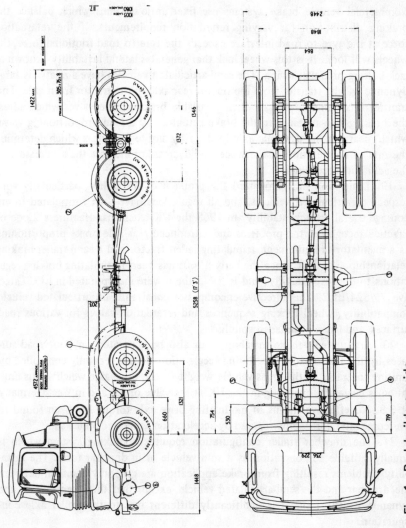

Fig. 1. Bodybuilders guide

accelerating at maximum tractive effort produces a considerable transfer of load from the front axle to the driving axle. This improves the tyre to road adhesion utilization but imposes restrictions on the steering stability.

18. Stability whilst braking a semi-articulated vehicle can create problems. Commercial vehicle brake systems use fixed ratio systems which balance the braking distribution for varying retardation requirements. If the retardation force at the tyre to road interface exceeds the tyre to road frictional force, the wheels will lock. It is this wheel lock that generates lateral instability—known as jack-knifing or trailer swing. The semi-articulated vehicles have a relatively large dynamic weight transfer from the tractor rear axle to the tractor front axle. This transfer is accommodated by fitting variable braking ratio valves which adjust the braking line pressure to the brake actuators by a load proportioning valve, which measures the dynamic axle load, or by anti-lock devices which determine the rate of deceleration of the wheel and decrease or increase the air pressure to the actuators accordingly.

19. Legislation has recognized the problem of instability, particularly with respect to articulated vehicles, and all regulations have been formulated to encourage overall vehicle stability. In 1968 the UK truck manufacturers' Code of Practice recognized the problems and introduced rear axle brake proportioning as a mandatory requirement, stipulating laden tractor and a semi-trailer braking relationship. The EEC since the early 1960s has been formulating braking regulations to improve stability, and in 1975 these were incorporated in EEC Directive 75/524/EEC. This directive encompasses constraints on articulated vehicle compatibility, wheel locking sequences and retardation values for various road surfaces, and laden or unladen conditions.

20. Articulated vehicle instability can also be generated on poor road surfaces by the loss of adhesion during acceleration. This is generally controlled by limiting the ratio of the gross vehicle weight to the drive axle weight. This limitation is currently operating in the UK at the value of 3.2 and in West Germany at 3.8. No effective means of modulating drive axle torque has been found to overcome lateral instability caused by acceleration generated wheel slip.

21. The drawbar trailer configuration requires a braking system which is equally suitable for operation as a solo vehicle or in drawbar form. The instability problems resulting from brake application are generally common between the drawbar and the semi-articulated vehicle, except that the magnitude of the dynamic weight transfer is significantly different due to the trailer axle configuration.

Speed

22. The problems of power, acceleration, and retardation are compounded

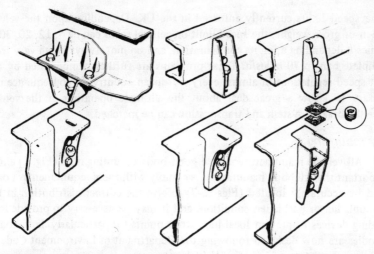

Fig. 2. Typical body mounting brackets

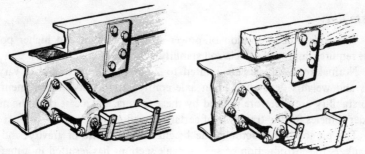

Fig. 3. Use of fish plates

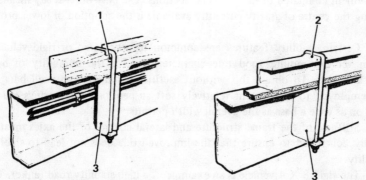

Fig. 4. Recommended methods of U bolt fixings: 1. steel capping; 2. shaped spacer; 3. metal spacer

by the speed limits currently enforced in the UK. Depending upon the vehicle unladen or gross weight, the speed limit on general roads can be 5, 12, 20, 30 or 40 miles/h depending upon its configuration, and on motorways it can vary from a complete ban to 70 miles/h. These problems are further compounded by different speed regulations in almost every European country. In consequence it is necessary to know a great deal about the ultimate operation of the vehicle before the braking system and transmission can be matched.

Body mounting

23. All vehicle manufacturers issue good body mounting rules (Fig. 1) and it is important that all body manufacturers comply with these requirements. Loads should be accurately located (Figs. 2-7) to give the correct distribution in the laden, unladen or part-laden condition, and it may be necessary to provide load spreading devices to prevent local high stress points for particularly dense loads. All bodies are now designed following the Department of Environment Code of Practice on the Safety of Loads on Vehicles.[2]

Roll stability

24. The widespread adoption of power steering systems and higher power engines requires improved vehicle roll stability.

25. Normal road vehicles are limited to 2.5m overall width and to varying lengths and weights depending upon axle configuration. Any improvements in vehicle stability are therefore limited by these factors, and a case could be made for wider vehicles for the transport of certain hazardous materials.

26. The predominant factor is the height of the centre of gravity, and for road tankers the introduction of special tank sections has resulted in a marked improvement compared to early circular sections. The only other ready means of lowering the centre of gravity currently available is the adoption of lower profile tyres.

27. Certain stability features are common to articulated or rigid vehicles. Modern vehicles require good ride characteristics for the durability of both driver and load. To obtain this without sacrificing stability, anti-roll bars are being employed to supplement relatively soft suspensions, and springs are installed on as wide a base as the overall width permits.

28. Stiffness of the frame structure and lateral control of the axles must be carefully controlled to ensure that the improved ride does not lead to steering instability.

29. The rigid 8 × 4 vehicle is an example of a high stability road tanker. The long wheelbase and two-spring high rate rear suspension give good ride characteristics and the tank can be mounted very low. The tank itself lends rigidity to

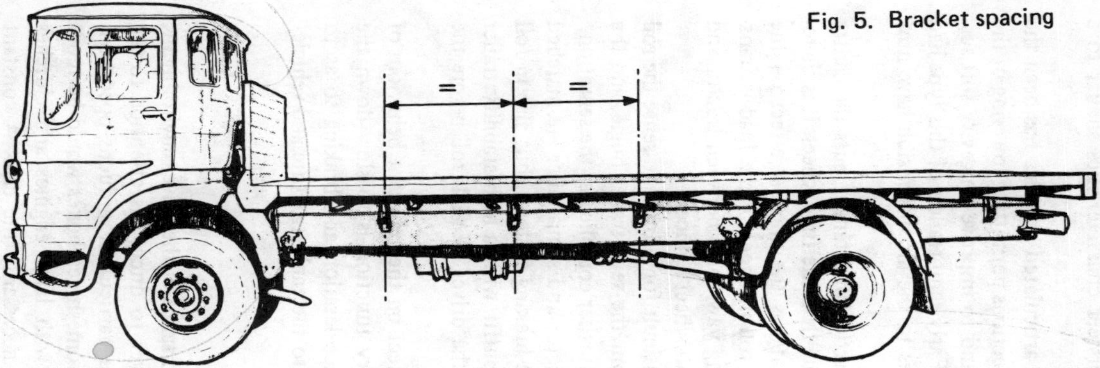

Fig. 5. Bracket spacing

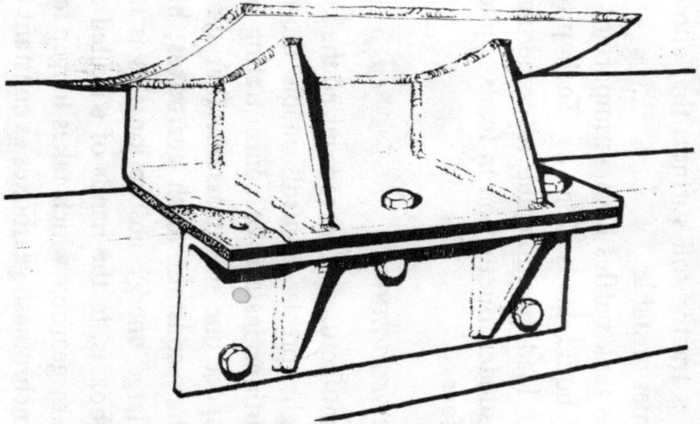

Fig. 6 (above). Typical body mount with flexible packing

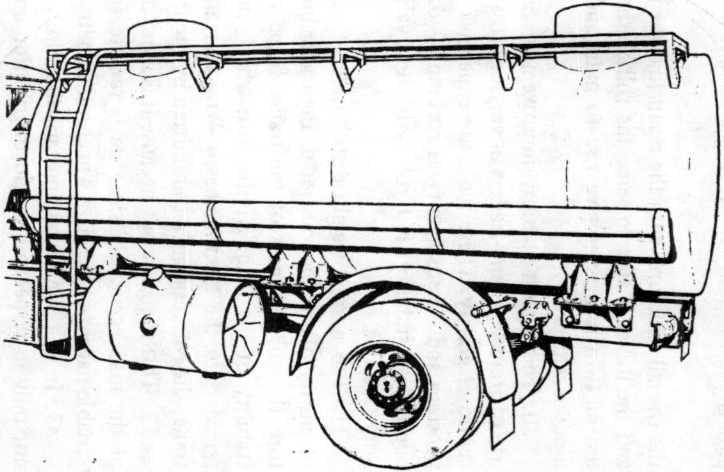

Fig. 7 (right). Spacing of tank mounts

the overall structure, with the result that the roll angles front and rear are similar and the driver is able to sense the roll of the load. The weight distribution and use of twin rear tyres give rise to mild understeer characteristics similar to a modern saloon car.

30. The biggest stability improvement on articulated vehicles has been the use of frameless semi-trailers of varying cross-sections permitting lowering of the centre of gravity. These tanks are expensive, and to improve longevity soft suspensions and super single tyres are employed. Early semi-trailers of this type had poorer roll stability, but roll restriction valves in the air suspension have now improved this.

31. The normal weight distribution of an articulated tanker puts the centre of gravity of the load nearer to the trailer bogie than the fifth wheel. It is therefore the roll stiffness of the trailer bogie that has the greatest effect on the stability of the load. At the fifth wheel the roll moment of the load is transferred onto the tractor chassis. With increasing yaw angle between tractor and trailer the roll resistance transmitted by the fifth wheel reduces.

32. This is significant because it is more difficult for a driver to sense the roll of the load accurately since he is remote from the semi-trailer bogie, and the variable nature of the fifth wheel roll stability further complicates the sensation.

33. It is, therefore, important that the tractor and semi-trailer are matched not only in respect of weight distribution and braking systems, but also in roll stability. The most desirable situation is an outfit with a high stability trailer bogie and a tractor roll stiffness that allows the driver to 'feel' roll before the load becomes unstable.

34. The tank baffles have an important effect on the transient behaviour of part-laden liquid loads (Fig. 8). Their primary function is to slow down the changes of load position due to cornering, acceleration and braking so as to avoid the sudden fluctuation in loads applied to the tank structure and vehicle/road interface.

Driver environment

35. Manufacturers have recognized the importance of the driver environment, because light, precise controls enable the driver to direct his attention to the road conditions instead of concentrating on a particularly arduous operation. In this respect one might assume that synchromesh gearboxes would be more popular than constant mesh gearboxes, but with large engines and clutches requiring large energy absorption this is not necessarily the case. A constant mesh gearbox is, in the hands of a skilled driver, much easier to handle than a synchromesh gearbox which takes longer to engage. Indeed many skilled drivers operate synchromesh gearboxes as constant mesh gearboxes.

Fig. 8. Effect on bulk liquid loads when negotiating bend

36. Automatic transmissions or semi-automatic transmissions can be expected to assume greater importance, especially where the movement of hazardous or abnormal loads is concerned, since in these instances it is even more important for the driver to concentrate his efforts on traffic obstacles. However, transmissions currently available generally suffer from a lack of range, and until this has been rectified and the number of gear steps increased, it is unlikely that automatic transmissions will supersede constant mesh transmissions. It is not considered that price alone is the only reason for the very limited range of automatic gearbox options available.

Current and future requirements

37. Legislation is currently proposed or enforced in the UK and in Europe on the following:

vehicle noise	door latches and hinges
emissions	audible warning devices
fuel tanks	rearward vision
underride protection	braking systems
rear registration plates	radio transmission
steering devices	interior fittings

- anti-theft devices
- seat strength
- external projections
- reverse gears
- speedometers
- statutory plates
- seat belt anchorages
- tachographs
- installation of lighting and light signalling devices
- side, rear and stop lamps
- reflex reflectors
- foglamps
- direction indicators
- rear registration plate illumination
- headlamps and bulbs
- towing hooks
- weights and dimensions
- safety glass
- head restraints
- windscreen wash and wipe systems
- windscreen demist and defrost systems
- identification of controls and indicators, and telltales.

Much of this information is currently of a conflicting nature and all has to be studied during the design phase of a vehicle.

38. The carriage of petroleum and spirit is controlled by special statutory requirements. Exhaust spark arrestors, firescreens and fully insulated electrical connections in flameproof boxes have to be fitted and Home Office approval sought for each vehicle.

39. In order to provide safer, more reliable, better tested, standardized vehicles, it is necessary that all this legislation and proposed legislation is agreed and unified within UK and Europe. It should, where possible, be based on performance criteria rather than design criteria, enabling individual companies to apply their ingenuity to devising safer, more durable trucks; not on specifying criteria which oblige vehicle manufacturers to fit more and more complicated proprietary equipment which, whilst overcoming the specific performance problems, also leads to significant losses in reliability and durability due to the increased complication.

References

1. IMechE/SMMT. *International vehicle legislation: order or chaos.* Joint Symposium, Automobile Division, Institution of Mechanical Engineers and Society of Motor Manufacturers and Traders, Institution of Mechanical Engineers, London, CP19, 1974.
2. DoE. *Code of Practice: Safety of loads on vehicles*, HMSO, London.

4. Design of tank containers
C. O. FARMER, W. P. Butterfield (Engineers) Ltd

Improvements in technology, particularly over recent years, and rapid increases in world demand have brought about a significant increase in the volume of categorized liquid dangerous goods requiring bulk movement, and tank container design has become a consideration of major importance. The tank container is designed to form an integral part with a modular framework, which conforms to the requirements of the International Standard Organization (ISO). The main considerations in the formulation of the unit design are the appropriate codes of practice, the selection of the materials to be used in the construction, the basic design parameters and production methods. The manufacturer must ensure compliance with and obtain approval from all regulatory authorities concerned, and must comply with the specific regional requirements which apply in the various countries in which the container is to be used. The experience gained in the design and construction of tank containers since their introduction has ensured that units can be offered to meet ever increasing demands, though problems of diverse regulations and requirements still exist.

1. The idea of a portable unit for worldwide movement of bulk liquids, whether hazardous or non-hazardous, was conceived and introduced in the form of the tank container some 10 years ago. Many changes have taken place in both the design and operational requirements, and much of the legislation and various codes of practice will continue to be reviewed. It is, however, relatively straightforward to design and construct tank containers suitable for the carriage of hazardous liquids, including liquefied gases, by road, rail and sea.

Design standards, codes and regulations

2. Tank containers must comply with and be constructed in accordance with the following codes and authority requirements, although the specific legislation will depend upon the area of unit operation. Typical standards are:

   (*a*) British Standard for Pressure Vessels (BS 1500 Parts 1 and 3);
   (*b*) American Standard Specification (ASME Section VIII, Div. 1);
   (*c*) the provisions of ADR (European Agreement concerning the International Carriage of Dangerous Goods by Road) applicable to tank containers;

Transport of Hazardous Materials, ICE, London.

(d) the provisions of RID (International Regulations concerning the Carriage of Dangerous Goods by Rail) applicable to tank containers;
(e) the provisions of the Inter-Governmental Maritime Consultative Organizations, International Maritime Dangerous Goods Code (IMDG Code);
(f) the report of the Standing Advisory Committee, Carriage of Dangerous Goods in Ships (the 'Blue Book'), and Annex 1 to the 'Blue Book' (Department of Trade);
(g) the requirements of the United States Department of Transportation (DOT) Highway, Railroad and Coast Guard;
(h) the relevant ISO requirements (contained in the International Organization for Standardization Document 1496, Part II: 1974 (3) Series 1, Freight Containers Specification and Testing Part III. (Tank Containers for Liquids and Gases)).

Types of tank container

3. A portable tank container (with the exception of type 5) means a tank that has a capacity of 450 litres or more, intended for the transport of dangerous liquids in bulk, with a vapour pressure of less than 2.94 bar absolute at a temperature of $50°C$.

4. Type 1 tank containers are fitted with relief devices and have a working pressure of not less than 1.724 bar. They may be used for numerous dangerous liquids, such as benzene, butyl acrylate, chlorophenols, ethyl acetone, phosphorous acid, etc., provided the pressures and materials of construction are adequate.

5. Type 2 tank containers are fitted with relief devices and have a working pressure of less than 1.724 bar. These tanks are intended for use with specifically listed less hazardous liquids, such as alcohols, butanol, camphor oil, ethanol, heptane, toluene, etc.

6. Type 3 tank containers are tanks without relief devices and have a capacity of 450 l or more, for the transport of certain dangerous bulk liquids with a vapour pressure of less than 2.966 bar absolute at a temperature of $50°C$. However, new construction of tanks of this type will not be authorized and the use of existing tank containers may only continue in service until 1 January 1980. Typical products for which they are used are acrylonitrile, methyl cyanide, toluene diisocyanate and xylyl bromide.

7. Road tank vehicles and type 4 tanks are not covered in this Paper.

8. Type 5 tank containers are tanks having a maximum allowable working pressure of not less than 6.897 bar, and having a capacity of over 1000 l for the transport of bulk liquefied gases with a vapour pressure in excess of 2.966 bar absolute at a temperature of $50°C$. They can contain such products as ethyl chloride, anhydrous mono-methylamine, butadiene, sulphur dioxide, etc.

Design considerations
Materials

9. Consideration in design must first be given to the products that are to be carried to determine the material required for the manufacture of the vessel. Account must also be taken of the product's specific gravity, and of temperatures and pressures, together with the unit operating areas and conditions.

10. In choosing the construction material, it is important to ensure that all necessary data on the product are obtained from the specific chemical manufacturer, for many hazardous liquids have detrimental effects on materials at differing temperatures and concentrations.

11. Materials used may be carbon steels, stainless steels, aluminium and nickel and nickel alloys, although the majority of units in present use are manufactured from the range of stainless steels.

Material thickness

12. Allowance should be made when determining the tank thickness for any corrosion likely to be encountered with a specific product and also any loading due to superimposed loads by the framework etc. Consideration should also be given to the calculated stresses, for these should not exceed 75% of the specified maximum yield stress or 37.5% of the minimum tensile stress, whichever is the lower, during hydraulic test. Minimum thicknesses are specified for the tank shell.

13. Type 1 tanks with a diameter up to and including 1.8 m should be not less than 5 mm thick if of mild steel or the equivalent if of other metal. Where diameters exceed 1.8 m the thickness should not be less than 6 mm of mild steel, 5 mm of stainless steel, or the equivalent in any other metal.

14. For type 2 tanks the minimum shell thickness should be as the type 1 tank except where adequate protection against damage is provided, this protection being in the form either of insulation acting as a 'sandwich' between the tank and outer sheeting, or of additional framework. However, the resulting thicknesses should not be less than 3 mm for tanks with a diameter up to and including 1.8 m.

15. For tanks with a diameter in excess of 1.8 m the minimum thickness should be 4 mm of mild steel or an equivalent thickness (but in no case less than 3 mm) of other metals.

Frame

16. With regard to the modular framework, one finds that two basic designs have emerged, the end frame configuration or 'beam' type and the full frame configuration (Figs 1 and 2). Both must comply with the overall dimensions of

Table 1 ISO designation and dimensions

Designation	Height		Width		Length		Rating	
	Metric, mm	Imperial, ft and in.	Metric, mm	Imperial, ft and in.	Metric, mm	Imperial, ft and in.	Metric, kg	Imperial ton
1A	2435^{+3}_{-2}	$8.0^{+0}_{-3/16}$	2435^{+3}_{-2}	$8.0^{+0}_{-3/16}$	12190^{+2}_{-8}	$40.0^{+0}_{-3/8}$	30480	30
1B	2435^{+3}_{-2}	$8.0^{+0}_{-3/16}$	2435^{+3}_{-2}	$8.0^{+0}_{-3/16}$	9125^{+0}_{-10}	$29.11¼^{+0}_{-3/8}$	25400	25
1C	2435^{+3}_{-2}	$8.0^{+0}_{-3/16}$	2435^{+3}_{-2}	$8.0^{+0}_{-3/16}$	6055^{+3}_{-3}	$19.10½^{+0}_{-¼}$	20320	20
1D	2435^{+3}_{-2}	$8.0^{+0}_{-3/16}$	2435^{+3}_{-2}	$8.0^{+0}_{-3/16}$	2990^{+1}_{-4}	$9.9¾^{+0}_{-3/16}$	10160	10
1E	2100^{+0}_{-5}	$6.10½^{+3/16}_{-0}$	2100^{+0}_{-5}	$6.10½^{+3/16}_{-0}$	2400^{+0}_{-5}	$7.10½^{+0}_{-3/16}$	7110	7

Fig. 1. 'Beam' type tank container

Fig. 2. Full frame type tank container

standard freight containers and the tolerances on these dimensions are as shown in Table 1.

17. The ratings for containers having these dimensions are also shown in Table 1. Although the container modular size designated '1C' which has a gross rating of 20 320 kg, thereby complying with the requirements of ISO 1496 Part III, is the one most generally used, more units of this size are being designated and manufactured for higher ratings up to and including 25 000 kg, thereby becoming non-standard ISO tanks.

Safety and relief valves

18. All tank containers with the exception of the type 3 must be fitted with a pressure relief device, situated in the vapour phase of the tank, this device usually being a spring-loaded valve.

19. In cases where the liquid or vapour is classified as highly toxic, harmful to the respiratory organs or entailing a poison risk, the pressure relief valve should be preceded by a burst disc assembly, to act as a product retainer. The space between the disc and the relief valve will be fitted with a pressure gauge to monitor the integrity of the disc.

20. The relief device should operate only under emergency conditions resulting from excessive rise in temperature, and since the tank will not be subject to undue fluctuations in pressure during normal operation, the relief valve should be set to start to discharge at a pressure 25% above the maximum allowable working pressure, and should reset 10% below the start to discharge pressure.

21. The relief device must be of adequate size to cope with conditions of fire engulfment as defined in Appendix 1, and have capacity sufficient to prevent a total internal pressure in excess of the test pressure of the tank.

Vessel attachments

22. With regard to tank fittings, all openings except the pressure relief devices and inspection openings should be provided with manually operated stop valves located as near the shell as practicable. Where any filling or discharge connection is located in the liquid phase of the tank contents, this should be fitted with an internal valve capable of manual operation, together with a second suitable closure on the outlet side (e.g., an additional valve or blank flange).

23. In cases where liquid products are designated as of high hazard, such as high corrosives and toxics, the vessels are not permitted to have openings in the liquid phase of the tank and are therefore provided with top discharge arrangements.

24. Where tank insulation is required, the design and construction should be such that the insulation in no way impinges on the specified requirements or

interferes with the proper function of the tank fittings. Due regard must also be given to the insulation material to be used to ensure that its capabilities match those of the tank's operating requirements. When heating provisions are required, due consideration should be given to the safety of the tank and its contents, since the development of excessive temperatures and stresses must be avoided.

Manufacture

25. Manufacturing tolerances are very important and the tolerances in plate rolling and the manufacture of the dished ends must conform to the appropriate manufacturing code. Regular checks of these tolerances are made during the various stages of fabrication to ensure that the end product is of the highest standard.

26. Most plates above a thickness of 5 mm are edge planed to ensure that full weld penetration is obtained. Cleanliness in this area is important to produce contamination free welds. Certain vessels require full X-ray on the longitudinal and circumferential seams to ensure that full penetration- and contamination-free welds have been achieved. Carbon steel vessels conforming to Class 1 requirements also require that a stress relief heat treatment be carried out to ensure that all stresses introduced during manufacture are removed.

27. The internal finish of the vessel is also of importance, since differing chemicals may be carried on consecutive trips after steam or detergent cleaning has been completed. Welds may be finished in varying ways; however, to ensure contamination-free loads the welds are usually dressed smooth. In instances where food products are to be carried, the internal surfaces of a stainless steel tank would be polished to give a bacteria-free finish.

Inspection, test and certification

Testing and inspection

28. All tank containers should be built under the survey of an approved inspecting authority and are subject to various tests and inspections as necessary during manufacture.

29. Full design details, together with calculations and other relevant data, are submitted to the appropriate inspecting authority and other regulatory bodies concerned, and must be accepted by them prior to manufacture of the tank unit. Specification for the materials used in manufacture and the standard of workmanship are both subject to examination and approval.

30. Test certificates showing the chemical analysis and mechanical properties of the materials, and the hydraulic test and radiography results, if applicable, would also be supplied, and in the case of new designs a test of a prototype tank

Fig. 3. Prototype test for bottom lift

container is required. Upon the satisfactory completion of tests, a type approval certificate is issued, certifying that the container showed no permanent deformation or abnormality which would render it unsuitable for use.

31. Normal tests are:
    (a) stacking strength
    (b) top lift
    (c) bottom lift (Fig. 3)
    (d) restraint
    (e) longitudinal inertia
    (f) lateral inertia
    (g) transverse rigidity (Fig. 4)
    (h) longitudinal rigidity.

Marking and certification

32. Each tank container must be provided with a permanently attached plate giving details of its manufacture, comprising manufacturer's name, date of manufacture, tank manufacturer's serial number, maximum allowable working pressure (bar), test pressure (bar), total water capacity (l), maximum weight of liquid to be carried (kg), maximum gross weight (kg), reference temperature zone, the control or permit identification of the approved inspecting authority, the hydraulic test date, the authority who witnessed the hydraulic test, codes, rules or regulations (by name or other identification) under which the tank is designed and the IMCO type.

33. Certificates are also issued for each tank container, recording similar particulars, together with a listing of the products approved for carriage and any comments regarding limitations to areas or type of operation.

Problem areas

34. Many users now have facilities for handling containers for international movement of many varied liquids. Containers can be hired or leased from various leasing companies, or the end users can purchase direct from the manufacturer. Many hauliers have turned to this method of moving liquids to most continents and are capturing a market previously serviced by drums and cylinders, by moving volume and therefore offering attractive rates. International chemical companies are the main users of hazardous liquids containers transporting by road, rail and sea to all parts of the world.

Fig. 4. Prototype test for transverse rigidity

35. Although every effort is made by all concerned to ensure that varying loads can be moved without contamination, problems do arise and are dealt with accordingly. Most of the countries concerned have equipment for the safe handling and movement, though certain countries have their problems in handling and the containers become prone to damage. Expense in repair is usually caused through mishandling.

Future development

36. The future in design, with an increasing demand, may see development in the use of higher tensile material to increase payloads and increase strengths. Improved methods and techniques in manufacturing and testing can be achieved as requirements for more and more various liquids are introduced to this type of transportation.

Conclusion

37. With the ever-increasing demand for the container tank to move products to almost every corner of the world, it has now established itself as a safe, economical and efficient way of moving many liquids by road, rail and sea.

38. Techniques have been developed through demand to give industries more detailed design, construction and testing, together with codes of practice that have made safer movement for all users. It is expected that the increase in demand will continue over the next 10 years and any changes in future legislation will automatically be incorporated in tank container design.

Appendix 1

[The following information is taken from the Department of Trade *Code for portable tanks and road tank vehicles for the carriage of dangerous goods in ships* (the 'Blue Book'): Annexe 1, 1977, and is reproduced by permission of the controller, Her Majesty's Stationery Office.]

39. The requirement for pressure relief under total fire engulfment conditions shall be considered to be complied with if the total air discharge capacity of the relief devices is at least equal to the value given by the following formula or if desired in lieu of the calculation, in accordance with the Tables.

$$F = \frac{8U(649-t)}{93580} \quad \text{(metric measures)}$$

$$F = \frac{8U(1200-t)}{34500} \quad \text{(imperial measures)}$$

where U is the overall thermal conductance of the insulation determined at 37.8°C (100°F) the units being kcals/m² h °C or Btu/sq.ft h °F.

t = the actual temperature of the substance at loading in °C or °F.

The value of F must in no instance be taken as less than 0.25.

Relief valve discharge capacity

Surface area, m^2	Minimum free air, m^3/h	Surface area, m^2	Minimum free air, m^3/h
2	841	37.5	9306
3	1172	40	9810
4	1485	42.5	10308
5	1783	45	10806
6	2069	47.5	11392
7	2348	50	11778
8	2621	52.5	12258
9	2821	55	12732
10	3146	57.5	13206
12	3655	60	13674
14	4146	62.5	14142
16	4625	65	14604
18	5092	67.5	15066
20	5556	70	15516
22.5	6120	75	16422
25	6672	80	17316
27.5	7212	85	18198
30	7746	90	19074
32.5	8268	95	19938
35	8790	100	20790

Imperial

Surface area, sq. ft	Minimum free air, cu. ft/h	Surface area, sq. ft	Minimum free air, cu. ft/h
20	28000	275	239800
30	39000	300	257500
40	49400	350	292200
50	59300	400	326000
60	68800	450	359100
70	78100	500	391500
80	87200	550	423300
90	96000	600	454600
100	104600	650	485400
120	121500	700	515800
140	137900	750	545900
160	153800	800	575500
180	169400	850	604900
200	184700	900	633900
225	203400	950	662600
250	221800	1000	691100

40. The heat input to the surface of the tank under fire engulfment conditions may as a minimum be taken to be:
$$H = 61015\, A^{0.82} F \text{ kcal/h}$$
$$= 34500\, A^{0.82}\, F \text{ Btu/h}$$
where
H = heat input to tank
A = tank surface area (m² or ft²)
F = insulation factor the value of which for non-insulated tanks to be taken as 1.0.

Where it is shown that fitted insulation reduces the heat input to the tank, credit for this may be given and F calculated as follows:

41. The total minimum certified rating of the relief devices would be:
Metric:
$$Q = \frac{10820 A^{0.82} F}{LC} \sqrt{\left(\frac{ZT}{M}\right)} \text{ m}^3 \text{ of air at } 15.6°C \text{ and } 1.013 \text{ bar}$$

Imperial:
$$Q = \frac{73040 A^{0.82} F}{LC} \sqrt{\left(\frac{ZT}{M}\right)} \text{ cu. ft of air at } 60°F \text{ and } 14.7 \text{ lb/sq. in a}$$
where
Q = required air discharge capacity/h (m³ or cu. ft).
A = external surface area of tank (m² or sq. ft).
L = latent heat of evaporation of the liquid (kcal/kg or Btu/lb) at relieving conditions.
Z = compressibility factor of vapour (if unknown a value $Z = 1.0$ should be used).
T = absolute temperature at relieving conditions (°K or °R; °K = 273 + °C; °R = 460 + °F).
M = molecular weight of the vapour.
C = a constant based on the rates of specific heats ($K = C_p/C_v$) of the vapour (a value of 0.606 corresponding to a ratio of 1.0 should be used in the absence of definite data).

Discussion: Papers 3 and 4

Mr P. Rundall (Air Products Ltd)
I represent an operator, a manufacturer of hazardous materials and also a tank manufacturer. There seems now to be enough legislation, and I think the time has come to contemplate the existing legislation and make the best use of it. I was disturbed by the notes on the Hazchem sign. Let's start using these signs now, rather than trying to extend them still further.

2. Because of the diversity of regulations, coping with operating problems and complying with regulations are almost impossible today. Over the last few years there have been tremendous increases in operating costs and it is a considerable problem to comply with all the regulations and still manage to transport goods at a reasonable and acceptable cost. Paper 4 dealt with tank containers, but tanks with pairs of wheels at each end are also governed by the regulations of the Department of the Environment, Department of Trade and Industry and ADR.

3. A point I should like to raise is that of vehicle stability. I recently attended a demonstration of a new series of trucks, but these were all flat bed trailers with loads of concrete placed with the centre of gravity as low as possible, and I am afraid this is the current trend. I would like Mr Cernes to comment on whether his company is doing something to change that situation. I have worked with British Leyland and understand the difficulties of trying to design a truck to deal with the various loading and unloading conditions and the variations in centres of gravity, but for tank containers, which have a high centre of gravity, this is an increasing problem.

4. In this connection driver environment has been referred to. British manufacturers are having to construct their vehicles so that they comply with European standards, with cabs which keep the noise down to 80 dB or even lower, and with fully suspended cabs. These isolate the driver from the load behind him. Most drivers will say that they drive by the seat of their pants, and today they cannot do that because they do not know what is happening.

5. I should like to know what is being done to ensure that vehicles are tested with high centre of gravity loads as well as low.

Mr L. Marsden, Tyne and Wear County Council
Paper 4 is a welcome review of design and operating criteria. However, the problem areas have not really been defined except to state that the containers become prone to damage.

7. My own observations over the years at the Tyne Tunnel are that these portable tank containers tend to be unstable in side surge conditions. I have seen

DISCUSSION: PAPERS 3 AND 4

at least two come off flat articulated trailers, with subsequent fracture of parts and release of the liquid, usually lager. Again, several have been noted with the relief valve blowing off, perhaps due to generation of gases under movement conditions. Again, the load was lager. Are these suitable containers for the transport of hazardous materials?

8. Based on the calculations in Table 2 of Paper 4, an IC tank 8 ft in diameter and 20 ft long produces 454 600 cu. ft of free air per hour relief valve discharge capacity. The discharge/s is therefore 125 cu. ft. Assuming still conditions, V cu. ft. of box shape, contamination all contained in the volume, uniform pollution, continual spread, and specific gravity of contaminant approximating to that of air, a table of contamination at various distances from the incident can be calculated (Table 1).

9. I would cite an accident report of the USA National Transport Safety Board[1] concerning an articulated tractor-tank semitrailer carrying anhydrous ammonia, in Houston, Texas. This accident happened on a bright sunny day, with temperatures of about 80–85°F and a 7 mile/h wind. There was a complete rupture of the tank. The released ammonia vaporized and the wind moved it downwind about 2000 ft until its effects were minimal. Within 3 min. of release, the maximum width of the vapour cloud over the ground was 100 ft, dispersing in 5 min. The accident statistics are given in Table 2. The injuries indicate the effects of an accumulated dosage over a period of 2–5 min. Because all fatalities occurred within 200 ft of the estimated release point it was assumed

Table 1. Contamination

Distance from incident, ft	Volume of envelope at specified distance from tank above ground, cu. ft	Pollution after 1 s, ppm (by vol)
1	1 980	63 131
10	27 360	4 569
100	4 942 080	25
1000	4 088 609 280	0.03

Table 2. Accident fatalities from rupture of ammonia tank

	Drivers	Passengers	Other
Fatal injuries	4	2	0
Non-fatal injuries	20	12	146

that within this distance the ammonia concentration was greater than 6500 ppm for at least 2 min. Concentrations of 5000-10 000 ppm, the report stated, are rapidly fatal for short exposures.

10. The report further gave calculations on the lateral liquid surge, and assumed from evidence that the speed of the vehicle was 53.6 mile/h into a curve. The calculated centrifugal force was 29 458 lb. The centre of gravity height was calculated to be 68.5 in. above the road. With demountable tanks this would of course be much more than 68.5 in. At that height the lateral force necessary to overturn the vehicle at 53.6 mile/h was calculated as 49 000 lb: 49 000-29 000 = 20 000 lb, an additional side surge figure, 70% of the centrifugal force due to solid loads.

11. The problem areas by definition are as follows.

(a) Portable tanks are not adequately baffled as with designed fixed tanks, though it is not just a question of baffles or speed. I myself have seen what one would assume to be an adequately baffled portable tank—a complete load of 2 gal. rectangular containers of washing-up liquid—which gently turned on its side at a slow walking pace when being driven round a roundabout, without even the wing mirrors being damaged.

(b) The centre of gravity is high because the tank is placed in a frame and on a platform.

(c) The fixings are less secure because the tanks are temporarily placed by a driver, who relies on the deadweight of the tank, which is not subject to the Construction and Use Regulations.

(d) While the petroleum companies and chemical manufacturers have constructive training programmes for drivers, this is not the case for drivers of flat-topped vehicles. Training courses for drivers of these do not necessarily include liquid loads and surge problems.

12. How can these problems be overcome to the degree that loose tanks on vehicles become at least as relatively safe as integrally designed tanks? I have never seen an integrally designed tank turn over because of side surge problems, though I have seen them turn over because of driver problems. I have seen only one case of a petrol tanker with a leaking valve. In general the companies adopt a responsible safety attitude to the maintenance of these designed tanks.

Mr H. D. Peake (Director of Technical Services, London Borough of Ealing)
Referring to my contribution to the earlier discussion, I would ask that the system of design of petrol tankers be re-examined to ensure that if one falls over, the compartmentalization does not fail.

DISCUSSION: PAPERS 3 AND 4

Dipl.-Chem. W. Hopfner (Bundesanstalt für Materialprüfung, Berlin)
In the development of higher tensile materials for tank containers I can see a problem because they are used to reduce the thickness of tanks and are more sensitive to stress corrosion cracking. Also in the case of accident they are very often not so ductile.

15. Because the thickness of tank containers is based only on the pressure of the substance and the yield strength of the material, it is possible to transport high-pressure substances in tank containers constructed of high tensile steels of very little thickness. However, careful investigation must be carried out before licensing.

16. Nobody wants to build tank containers of more than 3 mm thickness if possible, so manufacturers start with the absolute minimum thickness. On the other hand, there is usually a small corrosion rate.

17. I should like to know what happens when, after about 12 months or so, the thickness of the container is only 2.8 mm? Is the container then discarded?

Dr S. J. Maddox (The Welding Institute)
The Welding Institute Research Laboratory is concerned with all aspects of welding technology. My particular interest is the service performance of welded structures, such as the road tankers used for transporting hazardous materials.

19. The presentations on the design of vehicles have indicated that, on the whole, there are few design problems, apart from the inconvenience of having to satisfy a variety of European regulations at the same time! Also, Mr Rundell suggested that there was too much legislation, implying that all aspects of design are well covered.

20. I found this puzzling, because my organization is frequently consulted on cracking problems in road tankers, usually due to fatigue, which is a major problem in welded structures. I think these problems arise because there is insufficient legislation covering the design and construction of certain major items, so that manufacturers who are not technically aware can unwittingly introduce the causes of the problems.

21. Although I can accept that the larger manufacturers of vehicles pay sufficient attention to design, many road tankers are produced by small companies who do not have access to or even knowledge of design standards. The result is that components may not be designed at all, and in particular they are not designed to cope with fatigue loading. For example, bought in axle assemblies and supervision units may be incorrectly welded or attachments made in highly stressed areas.

22. Many examples of failures have been associated with welded joints which anybody with the least knowledge of fatigue in welded structures would not

allow. Thus, guidance is needed on the choice of welded details. Also, design stresses are often found to be too high because they are based on some proportion, commonly two thirds, of yield strength. This is a meaningless concept in the context of fatigue of welded joints where failures can occur at stresses of 10% of the static strength.

23. One of the most thorough structural design standards is concerned with static vessels for holding dangerous fluids and fluids under pressure but there seem to be no standards applied to vehicles which might transport such vessels.

24. I believe there is a clear case for the introduction of new design rules covering the construction and service performance of welded vehicles carrying hazardous fluids. The time is particularly ripe in relation to fatigue aspects because over the last few years the design of welded joints has been closely examined and major revisions are currently being made to existing standards covering fatigue in welded structures, the main one being the steel bridge design standard BS153. Unless the people concerned with buying or making vehicles for carrying hazardous materials are aware of these standards and are obliged to make use of them, this information will not be used in a highly relevant area.

25. Finally, with reference to Dr Höpfner's point about the use of high tensile steels, it should be noted there is no difference between the fatigue strength of welded high and low strength steel. Therefore, if high tensile material is used to allow an increase in design stresses, the risk of fatigue failure increases.

Mr W. McMillan (Director, P & O Road Services)
I have been in the bulk liquid transport industry for some 25 years and in that period quite a few of our vehicles have overturned. When a vehicle overturns it has forward momentum, and in my experience it is this which causes the damage. In a tank wagon with no insulation the barrel takes the impact and the abrasion, and it is usually fractured or worn away. The product leaks and sometimes a fire follows. On two occasions insulated vehicles were overturned, and/despite bad crumpling of the barrel they were not pierced and the load was intact.

27. This in itself raises another problem. When a flat bed lorry or a vehicle carrying a container is overturned, the load is shed. The load can be picked up piecemeal, and the container is made for lifting. But a laden tanker is a most difficult thing to put back on its wheels or move to the side, and spills have resulted when recovery teams breached the tank during their efforts to right it. As a direct result of such experiences, the Road Haulage Association has arranged a load recovery scheme. The basic idea is that the accumulated knowledge of the industry should be used to supply a relief vehicle to receive the load and the know-how to transfer it from the vehicle involved in the incident to the relief vehicle.

DISCUSSION: PAPERS 3 AND 4

28. I should like to add that some of our vehicles have been in service for about 20 years and we have no experience of fatigue fractures. Presumably the methods used by manufacturers in the past have proved adequate in that respect.

Mr G. Mackay (Redland Purle Ltd)
On Mr Rundle's comment that there is sufficient legislation I would point out that so far as the law is concerned in Britain the most hazardous material can be carried in a polythene bag, provided it does not leak and it has sufficient labels on it. I think priorities are wrong if there is legislation, which I welcome, for labelling and no legislation for wrapping up the package.

30. A number of things need to be done. Surely the first thing is, under the Health and Safety at Work Act, to confine a load under all reasonable foreseeable circumstances. A roll-over is a foreseeable circumstance, and happens often enough. The petroleum industry must have four in a year. A petroleum tank is frequently constructed of 10 or 12 gauge material and tanks are still being built to that standard. In no way can that tank contain the load in a roll-over.

31. An earlier speaker referred to the problems of waste. There may be contractors who do not notify. There may be contractors who carry around waste without knowing what it is, but there are others who do know what they are carrying. There is a system of Tremcards for waste and a Hazchem coding for waste, but until these are published by government they cannot be used.

32. What is necessary is an independent inspecting body for road tankers. My firm uses a firm accustomed to boilers who inspect all tanks regularly. But their knowledge of chemical tankers is nil, because they regard a tanker as a boiler on wheels.

33. As a final question, should there be some legislation on minimum tank thickness, based on the principle of a pressure vessel?

Mr J. R. McIlwraith (Chemical Sales Dept., British Celanese Ltd)
Each day from my office I see large numbers of road tank barrels entering and leaving our factory, and in spite of my involvement in the area of chemicals movements, to my knowledge British Celanese have never been asked by a design company to supply details of the properties of our chemicals.

35. How do they get their design information therefore, and how is this extended to chemicals which are not classified in ADR, RID or UN systems, into which many of the chemicals produced by Celanese fall?

36. How does one therefore cope with legislation on tanker design when the regulations, although prolific, do not seem to concentrate on unlisted hazardous chemicals produced by UK manufacturers?

37. In my opinion large bulk UK produced chemicals should be easy to col-

DISCUSSION: PAPERS 3 AND 4

late, but certainly during the last 2 years since I have been dealing with this area, no designer of bulk road barrels or ISO containers has asked our advice.

38. As Britain is an island and any international shipments must comply with the Department of Trade 'Blue Book', I find this extremely surprising and would welcome advice on this area.

Mr M. C. Smyth (Chief Engineer, Dublin Port and Docks Board)
My authority has been involved in the importation, bulk storage and transportation of tankers containing liquid with a flash point of less than $-4°C$. One of the operational conditions which has been insisted upon is that the containers should not be filled above 93% of their capacity, and as well as pre-set meters, there are sensing devices in the containers which ensure that the pump will be automatically cut out if this capacity is exceeded.

40. On reading Paper 4, I saw no reference to the question of filling tank containers. Is this aspect considered in the design?

41. Mr Farmer states that the relief device should operate only under emergency conditions, and I wonder if that statement is absolutely correct. Depending on the method of filling the tank, temperature, pressure, etc., a tank container could emit hazardous vapour within non-emergency conditions.

42. Could Mr Farmer give details of the problems in handling in certain countries to which he refers. I have been concerned for some months with the question of handling containers both on despatch and arrival. From enquiries I have made in the UK and in Europe it would appear that tank containers containing very hazardous materials are handled by ordinary equipment such as van carriers and fork lift trucks. There appears to be no attempt to make these non-sparking. The petroleum industry insists that vehicles should be specially designed and constructed to handle hazardous materials. I wonder if Mr Farmer could comment.

Mr M. Pitt (Aston University Chemical Engineering Department; Technical Advisor to Fospur Ltd)
What does seem to cause problems in chemical tanks, and where a lot of experience is required is in the design of valves. This is a physical weak point which also requires human intervention. On road tankers it is obvious that there are both valves and couplings, with hoses to be joined together and unjoined. Filling and emptying are potentially hazardous operations, and there have been incidents on both static and mobile tanks where people have been splashed or large discharges have occurred, arising out of incorrect use and sometimes incorrect understanding.

44. I would be interested to have information on the current state of design,

DISCUSSION: PAPERS 3 AND 4

particularly on standardization of valves. I would like to know if there is a standard method of operation to ensure that a driver using an unfamiliar vehicle will use the correct procedure, and that in an emergency any individual dealing with valves or couplings does so correctly. It is easier to transfer contents from leaking tanks if fittings are compatible. On the other hand one can use coupling design to prevent people putting the wrong materials into a container, as in the use of right and left-hand threads on gas cylinders for example.

Mr E. A. George (Agricultural Division, Imperial Chemical Industries Ltd)
I recently became concerned with the design procurement of pressure vessels on wheels. A pressure vessel on wheels is neither just a vehicle, nor a boiler incidentally having wheels, because when it is connected for filling or emptying it is an integral part of the plant to which it is connected. In the UK the majority of incidents, particularly with ammonia, occur during the filling or unloading operation rather than on road or rail.

46. Matters such as couplings and valves and methods of emptying tankers are within the remit of the large chemical manufacturers and their people who produce the chemicals. ICI has done much research on the best type of hose or solid coupling and the best type of valve to use, and we have made sure that what is used for liquid CO_2, for example, cannot by any mischance be used for ammonia: mixing pressurized liquid CO_2 with pressurized ammonia is not very pleasant.

47. Another thing which appals me is to hear calls for more legislation or inspecting authorities. The draft HSE proposals for pressurized equipment, which includes transport pressure containers reflects the thinking of the manufacturing industry that it is our job to make sure that things are done properly and that things are fit for service and are, indeed, inspected. Certainly large companies have in-house inspection facilities which are truly independent of commercial pressures.

48. The other point I would like to make is on vehicle roll-over. A particular incident in America involved a tanker which was only about two-thirds filled. The driver was frantic to get home because his wife was expecting their first baby, and he drove the tanker round a corner far too fast. Professor Newland of the Engineering Department at Cambridge, who specializes in rail suspension stability, has suggested longitudinal baffles or filling the tanker with aluminium pan scrubbers. The Petrochemicals Division of ICI actually fits longitudinal baffles in its spirits tankers, and what were, in effect, aluminium pan scrubbers were originally conceived to protect fighter aircraft fuel tanks. These might also help to stabilize the road tanker.

DISCUSSION: PAPERS 3 AND 4

Mr B. Hooper (Pullman Kellog Ltd)
Further to Mr George's remarks on ammonia movement, it would appear that carbon steel is still used for the storage of liquid ammonia, although it is well known that this material suffers stress corrosion cracking after long periods. This is a subject which must be considered seriously in the near future. Possibly one way of overcoming this would be to ensure that the carbon steel is stress relieved, and this can be done for static storage tanks. One wonders if this remedy would work with mobile tanks which are subject to fluctuating stresses in operation.

50. There is a common fallacy that stainless steels are safe materials to use where corrosion is concerned. Austenitic stainless steels are notorious for developing stress corrosion cracking in hot caustic or chloride solutions, even in small amounts of chlorides, and it is sometimes difficult to eliminate the small amount of chlorides in many liquids carried.

51. The use of plastics has not been mentioned. Plastics, I think, have a certain application in the movement of dangerous materials. In India I saw a hydrochloric acid tanker which had overturned on the road. The tanker itself had been made by hand from glass fibre reinforced polyester, and was intact, though the lorry it had fallen from was a wreck.

52. Something else which has not been mentioned is the movement of liquid metals. I have not heard of any serious accident occurring on the roads, or even on the railways, but molten metal is moved distances of 20 miles regularly, and one wonders what would happen if a leak or some particular catastrophe occurred and one of these containers met with an accident in transit. I imagine fire-fighting people would not be particularly well prepared.

Mr W. E. Huddart (Hoechst UK Ltd)
Tanker corrosion is hazardous and can also stain or otherwise contaminate the cargo. The anti-corrosive strength of materials of construction used for the barrels of road tankers, and especially linings, is a constant problem to transport companies, vehicle manufacturers and chemical suppliers. In international traffic the return load may be unspecified and have different properties.

54. What type of research is being carried out by vehicle builders, bearing in mind that conditions inside a road tanker differ from those in a static tank? For example, in transit the load is in constant movement and the product will intermix with the gas in the inner atmosphere.

The Chairman (Sir Charles Pringle)
Jack-knifing is a matter which comes to the attention of the public a good deal. I should like to ask Mr Cernes how widespread is the use of Maxaret-type anti-

skid brakes. These have been well proven in aircraft and in other fields. Should they be a requirement for articulated vehicles?

Mr Cernes (*Paper 3*)

Mr Rundle's point about stability is very relevant, and raises a number of issues. There is no doubt that, over a period of years, a lot of the tractor design has been developed on the use of flat bed trailers, mainly because most of the large manufacturers do not own tanks and if they did it would be difficult to identify how representative they are of the pattern which is supplied commercially. Notwithstanding that there has been a move for some years now to use high centre gravity trailers, even these are not necessarily representative of fluid loads in every case. During a cornering manoeuvre there is not only a lateral movement of the load, but also a longitudinal component which tends to move the centre of gravity of the load forward in the tank. There is a triangle of stability between the fifth wheel and the trailer suspension, and this stability also varies with the articulation of the tractor and the trailer. In fact, a case can be made for turning the fifth wheel upside down and having the articulating fifth wheel on the trailer and the pin on the tractor. This would give a more constant roll stability at the front.

57. With regard to isolating the driver from his environment, my company and I are against too much isolation of the driver. We still believe that the design of the vehicle should be such that the driver can recognize instability in his trailer or his load. We do a fair amount of work in balancing the roll stability at the various axle stations down the vehicle. One of the problems is that flat bed trailers have a very low roll stiffness, and tankers have a very high roll stiffness. The task is to obtain a coupling which will transmit trailer instability irrespective of the yaw attitude of the tractor and the trailer, and the driver in his cab should roll rather more than any other part of his vehicle. As I mentioned in the introduction to my Paper, this is compounded by power steering so that the driver is likely to tackle roundabouts and corners with a good deal more enthusiasm than if he had to work hard at it.

58. Cab suspension is another point which raises some controversy. As we all know, one or two trucks have recently come out with full cab suspension. This is not universally applauded, on the basis that this introduces yet another system between the driver's seat and the load, and whilst it is satisfactory to have a suspension which will take care of the longitudinal movement of the vehicle, the cab suspension should not isolate the driver from lateral movement.

59. As a chassis manufacturer, the only comment that I have about the design of containers is that the centre of gravity should be as low as possible.

We have a $2\frac{1}{2}$ metre overall width, a fixed width of tyres in which to accommodate the imposed loads on the ground, therefore, the spring base can only be so wide. A case could be made for hazardous materials going over the $2\frac{1}{2}$ metres if there is a particular requirement for more stability where it is required to move reasonable distances in a reasonable time, provided that the route could be controlled. I would not suggest that one took a 3 metre wide truck through villages—there is no reason why 3 metre wide trucks should not trunk up and down some of the main motorways. As one who works for a company making 3.1 metre wide vehicles, and located in the middle of a town, I can appreciate the problems of getting them through.

60. The retention of containers on chassis is adequately covered by the Department of Environment Code of Practice; containers should be retained with twist locks. I do not have any sympathy with the company that operates with a ball of string and a piece of chalk.

61. Mr Maddox raised the question of the fatigue life. I was not too sure whether he was referring to the suppliers of the chassis or trailers when he was referring to the fatigue life of axles and suspension units. Certainly amongst the major chassis manufacturers fatigue is a criterion that is considered during the design of axles and suspensions, and indeed the whole vehicle. The problem is identifying the load; the frequency with which load variations are applied and their magnitude. Notwithstanding that, we have routine test procedures and these are correlated to operating experience. It is not possible in its own right to say that because it passes our standard it will satisfy every requirement. All we can say is that in our experience it has satisfied the majority of the requirements. If there is an unusual operation there must be close collaboration between the ultimate operators, the tank supplier and the chassis manufacturer.

62. Mr Mackay raised the question of independent inspectors. I have a good deal of sympathy with Mr George. I think it is far better that the manufacturing industry adopts, for instance, the Ministry of Defence 0521 standards, whereby the company and its procedures are assessed and audited rather than getting specific inspectors who after a while tend to get somewhat out of touch with commercial practice. One can end up, as one does with the passenger vehicle industry, when an independent arbiter, who is concerned with neither the design nor the operation, controls the standards.

63. Vehicle roll-over, to which Mr George referred, is a problem. Really one can only contain vehicle roll-over by containing the load. One can increase the roll stiffness of the vehicle, but the driver should sense the worst roll in the vehicle. One can go so far towards improving roll stiffness by fitting anti-roll bars but it is a question of balancing the chassis characteristics with trailer and load operation, and I think it can only be done effectively by the industry.

64. With regard to the Chairman's point about jack-knifing, it is a lot more complicated than anti-wheel locking devices. Basically the cause of articulated vehicle instability is wheel locking. Whether one prevents wheel locking by restricting the braking force applied to the axle in accordance with its dynamic load—and here there is quite a variation between the dynamic loads and the braking, acceleration, and uniform motion—or whether one applies it by sensing the rate of deceleration is relatively immaterial, although the cost difference is very significant. What matters with all these units is the maintenance that is applied to them. They all work. There are claims that merely by sensing the drive axle one can eliminate jack-knifing, and in the strict sense that is true, but one can get just as much lateral instability by locking non-driving axles. If the trailer is locked, it will swing round. If the front axle is locked, the front of the truck will follow the camber down into the guttering and the vehicle will jack-knife in that way. One of the problems in all the roll-over incidents, and all subsequent claims that the chassis was at fault and it was nothing to do with the driver, is identifying just what mode of failure occurred. Did the trailer swing first before it broke? Did the tractor front break away? Quite often the driver in the incident is the man least able to provide a cool assessment since he was panicking quite dramatically at the time. It is very difficult to quantify roll-over accidents. There are so many variations of them. All we can ask is that the anti-wheel lock devices that we provide are well maintained and regularly tested, and that the operators and the manufacturers move towards self-certification and self-control in the use of their equipment.

Mr Farmer (Paper 4)

The relief valve discharge capacity figures quoted by Mr Marsden appear to be accurate. We, as a manufacturer, work to codes of practice that are legislated by various countries. Additional safety factors are added for the carriage of some hazardous products, such as bursting discs and increased thickness in vessels.

66. In answer to §11.(a), many tanks that carry part loads are so designed as to include at least two baffles. The centre of gravity is higher than on a conventional tanker. However, taking into account the number of freight containers and liquid containers in current use, the number of accidents that have occurred are so few that the overall height and centre of gravity do not appear to be detrimental. Twist lock fixings have been developed for holding down all types of ISO containers

67. Regarding Mr Peake's contribution, over the years every effort has been made to ensure that tankers are made strong enough to withstand overturn conditions by maximizing each compartment to 1100 gallons.

68. Mr Hopfner mentioned the use of high tensile steels. In the majority of

cases, tanks are designed to take into consideration the maximum developed pressure experienced in the areas of operation. Generally speaking, high tensile materials are not used by UK Manufacturers. One British code stipulates a minimum thickness for tanks carrying high corrosives; this thickness being in excess of the minimum thickness determined by calculation basing on pressures. Should any container have experienced thinning due to corrosion, then one would expect this unit to be derated.

69. With regard to Mr Maddox's contribution concerning standards, as more and more codes of practice are introduced, design problems, in particular cracking, are eliminated through the use of thicker vessels and more substantial mountings and so on. With the additional legislations for the carriage of inflamables and corrosives, further problem areas should be resolved.

70. The only comment I have to make on Mr Mackay's remarks are that petroleum spirit tanks are designed to conform to design notes layed down jointly by the Health and Safety Executive, and by Industry.

71. To answer Mr McIlwraith, information on the properties of chemicals are obtained by manufacturers and if they are not covered by ADR, RID or UN they are referred to the DOE, for classification.

72. With reference to Mr Smythe's remarks on overfilling, certain tanks can and are fitted with a variety of loading gauges and alarm systems. When designing the tank, filling ratios are always taken into account. Tanks designed for liquids emitting hazardous vapours have venting systems to prevent omissions during normal operating conditions.

73. In some parts of the world containers are handled roughly, however, they are designed for handling in most conditions to safeguard accidents.

74. Mr Pitt mentioned standardization. The industry over many years has with some success, obtained standard fittings and couplings, however, much should be done to improve standardization.

75. With regard to corrosion as mentioned by Mr Hooper, generally most ammonia tanks are manufactured from carbon steel and may be subject to stress corrosion cracking after long periods. However, the majority of tanks are stress relieved to alleviate this problem.

76. Stainless steels are not always safe materials to use where corrosion is concerned and in many instances alternative, more suitable construction materials are used.

Reference

1. US National Transportation Safety Board. *Report of accident 11 May 1976.* NTSB-HAR-77-1, Washington.

5. Routing of hazardous substances moved by road

W. G. ASHTON, MBE, MBIM, Deputy Chief Constable, Cleveland Constabulary

The routing of vehicles carrying hazardous chemicals and petroleum products away from urban centres and other similar places has such advantages that, despite practical difficulties, it merits further study. The tonnage of such substances moved by road is considerable. Most of this tonnage is moved in bulk by chemical and petroleum tankers. About 600 different chemical substances are carried by road in bulk in the UK. Given an estimated 250 incidents per year concerning chemical/petroleum tankers, the need for some attempt at pre-emption of public risk exists. Shopping areas and road tunnels are obvious places to be avoided but there are other places of similar risk. Difficulties there are, but if the need for safety is willed does it not follow that the means should also be so willed?

Events during the past few years have focused considerable public interest onto the potential dangers of transporting hazardous petroleum and chemical products. It is an emotional subject on which discussion has prompted the notion of banning such movement, by road at least. On the premise that there are risks inherent in moving hazardous substances it is arguable that in the interests of public safety some constraints are necessary. To be effective, however, these constraints need to be realistic in economic as well as social terms.

2. More than 50% of all goods transported throughout the world are to some degree hazardous—especially those produced by the oil-based and chemicals industries. These industries supply essential raw materials to almost every other industry. As the world and domestic economies move to become more dependent on manufactured goods, this supply and therefore the volume of transport must increase.

3. The tonnage of such substances moved by road in the UK is considerable—about 34 million tons of chemicals and 85 million tons of petroleum products. But it is a small percentage (2% and 5% respectively) of the total tonnage of freight that goes by road. Most of this tonnage is moved in bulk by some 5000 chemical tankers and 11 000 petroleum tankers. There are, of course, general purpose platform vehicles in use carrying mixed loads of packaged goods of a multitude of chemicals and mixtures. In the latter case relevant constraints are less easy to devise but on the premise of hazard the case for them remains.

Transport of Hazardous Materials. ICE, London.

PAPER 5

What is a hazardous substance?

4. The definition of what is a dangerous or hazardous substance is difficult to formulate since in the final analysis almost any substance carried by road or rail can prove dangerous or hazardous depending on the circumstances of its spillage or unshipping. Generally speaking, however, the danger or hazard depends on the degree of toxicity, corrosiveness, explosive potential or flammability of the substance. These are, however, relatively crude terms in which to describe the wide range of hazards which may arise from the spillage of any one of a complex variety of substances. The most universally accepted definition is that prepared by the United Nations Committee of Experts in which each substance is given a coding and number based on its unique qualities (Fig. 1).

Movement of hazardous substances

5. The movement of such substances by road is not of itself hazardous if:
   (*a*) the vehicle is suitable, properly maintained and sensibly used;
   (*b*) the packaging arrangements are secure;
   (*c*) the substance carried is not incompatible with its packaging, with other substances carried on the same vehicle, or the immediate ambience of its carriage.

The hazards that arise do so from the pathological effect of the substance spilled in consequence of:

   (*a*) a road accident, or
   (*b*) a malfunction of equipment.

There is not, however, any evidence to show that drivers of vehicles carrying hazardous substances are any more liable to be involved in a road accident than drivers of other commercial vehicles. Equally the packaging arrangements and specialized vehicle and associated equipment used by responsible operators do not seem particularly prone to failure.

6. Because of the absence of an obligatory reporting and recording of incidents involving hazardous loads it is difficult to identify the size and shape of the overall picture. The best advice seems to suggest a figure of about 250–300 incidents per year. However, these are probably those requiring positive action and may not include 'scare' situations which misinformation or lack of information may produce.

7. There are, however, some substances which by reason of their particular hazard ought not to be moved by road or rail. It is difficult and probably not desirable to catalogue these substances by name. Nevertheless a clear proscription should apply if the substance, if spilled, could by reason of its immediate action cause death or serious malfunction of the human body.

Fig. 1. UN substance number

Preventative action

8. So far as the safety of the general public in streets or places nearby is concerned there are at least three main lines of preventative action.

   (a) the directing of vehicles carrying hazardous substances away from areas of high accident risk or confined spaces such as tunnels, etc., and other places where the public may congregate;
   (b) the clear, overt marking of vehicles carrying hazardous substances whereby in the event of fire or spillage, members of the public, vehicle drivers, fire brigade and police personnel are quickly made aware of the particular hazard.
   (c) the planning and provision of informed, effective advice and practical rescue/recovery facilities to be available quickly in the event of a spillage or unshipping.

9. Although the purpose of this Paper is to look at the first of these three lines of action, it will be helpful to refer to the other two items, since to be effective a route constraint scheme needs to be complemented by a hazard information system which embraces these items.

Hazard information system

10. The essential feature of such a system is that it must serve a number of purposes simultaneously:

   (a) to provide a warning of impending danger to life and to the environment
   (b) to enable immediate action to be taken to safeguard human life or to prevent further injury to human life;
   (c) to answer the questions 'what are the hazards?' and 'what should be done?'
   (d) to enable actions to be taken that will not be inconsistent with full remedial and hazard suppressive actions that can be taken later.

11. In this context it is important to note the various types of individual who can become involved in an emergency situation. 'Members of the public' for example could refer to anyone from a young student to an aging grandparent. Clearly their ability to respond will vary and in any event will be dissimilar to that of emergency services personnel, staff in chemical plants and drivers and loaders of vehicles engaged in the transportation of hazardous substances.

12. There are a number of hazard information systems in use throughout the world but only a few which will affect traffic in the UK. The principal system in use in this country and which is shortly to be supported in part by law is the United Kingdom Hazard Information System (see Paper 2, Fig. 3). The system currently in use on the European continent is based on an ADR requirement to which has been added a French originated code for identifying hazards (Fig. 2). Each system uses the UN substance number as a reference point; they differ in that the Hazchem code used in UKHIS is a prescription for action whereas the ADR/Kemler code is an informative system (i.e. properties code).

13. The UKHIS comprises three main elements:

(a) the carriage on vehicles of the UKHIS label and appropriate Tremcard: from the information on either of these documents the emergency services can take the right immediate action; additionally the label contains information on which subsequent action can be based;

(b) the provision by way of the CIA mutual aid scheme (Chemsafe) and the RHA recovery scheme of expert advice and recovery facilities: these schemes, which although provided by the two associations for their members are readily made available in incidents involving non-members, enable the emergency services to continue the immediate action and deal with the incident to its conclusion;

(c) the training of those personnel who are to be called upon to act in the event of an incident: Police, Fire and Ambulance service personnel have been, or are being, trained and exercised in hazard identification and rescue procedure; HGV drivers of tanker vehicles are being required to add to their driving skills an understanding of the substance hazard and what to do in an emergency situation.

14. It would be wrong to give the impression that all these elements are present in respect of each movement of a hazardous substance. This is not so since the labelling requirement yet requires the force of law and in any event will in the first instance relate only to bulk tanker loads. Package traffic marking is more complex and still has to be resolved. Other general aspects such as the compulsory carriage of instructions in writing, parking and the structural specifications of vehicles are similarly awaiting decision. Nevertheless a great deal has been done and UKHIS is a meaningful and useful system.

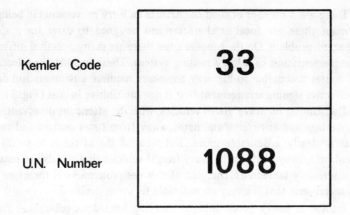

Fig. 2. ADR/Kemler label

Route constraint

15. The directing of hazardous substance traffic away from specific places of risk and the routing of such traffic generally involves similar but not wholly identical considerations. So far as general routing is concerned the case is much the same as for any heavy goods vehicle traffic. In this there are several factors at play; the damaging power of heavy lorries, the level of noise and the exhausting of diesel or petrol fumes. When to these factors is added that of environmental preservation there is adduced a strongly persuasive case for route constraints on heavy goods vehicles.

16. The principal argument against route constraints is that so far as the general distribution of freight is concerned there is no satisfactory alternative to the lorry. It has been suggested that the majority of freight movements should be switched to rail, pipeline, or inland waterways. The last has obvious drawbacks because of the restricted waterways network; in the case of the railways a severe limiting factor is the number of railheads. Movement of hazardous substances in bulk by pipeline has many advantages, particularly that of safety, but it does not seem to be practicable in servicing all customers or moving a variety of substances in relatively small amounts.

17. If, however, the lorry is to remain an important part of the goods distribution system and its nuisance effect is to be abated, there will need to be some form of route constraint on heavy goods vehicles. The aim is to make transport possible by eliminating risks or reducing them to a minimum. The problem is therefore one concerning safety no less than that of facilitating transport.

18. There are a number of such constraints on lorry movements in being, but in the main these are local in character and designed to cater for a specific traffic hazard problem. On the broader issue there are many practical difficulties in the implementation of a lorry routing system. These include the selection of suitable routes acceptable to highway engineers, hauliers, customers and drivers, and an effective signing arrangement that is not prohibitive in cost (Fig. 3).

19. The routing of heavy goods vehicles, with the attendant disadvantages of danger, damage and environmental harm, away from town centres and residential areas is clearly a desirable move. But what if the effect is to so fill main trunk and motorway routes with heavy freight vehicles that the risk to passenger traffic is seriously increased? The cost of this desirable move is therefore more than financial—and that is a very considerable factor in itself.

20. Insofar as heavy goods vehicles carrying hazardous substances are concerned, there are further considerations in relation to route constraint of which one, the risk appreciation factor, is critical and which usually produces a highly subjective judgment. Route selection of hazardous substance traffic calls for an accommodation of the following factors:

(a) *Risk appreciation.* The degree of risk is difficult to assess as this will vary enormously with the substance carried and the circumstances of the spillage. Nevertheless planning must go on and this can probably best be based on what experience has shown to be the more probable end results. The 'worst possible case' basis has some attraction but the factors are so variable as to bedevil contingency planning to a 'holocaust-every-time' concept.

(b) *Probability of an incident.* At best it seems that there will be 250–300 incidents/year but this may be a conservative estimate. In any event the probability will vary as the volume of hazardous substance traffic passing through any particular area varies.

(c) *Maintenance of commercial viability.* This is not meant to suggest that commercial viability alone should dictate or prohibit a choice of route. Nevertheless on the basis that hazardous substances must be moved by road, a balance has to be struck between absolute safety (which in any event is a somewhat abstract ideal) and what is probable.

(d) *Reasonable access by emergency services and rescue personnel.* Ideally this suggests a choice of roads with dual carriageways or which are complemented by available alternative routes. Practically, however, it will need to be seen in the way in which assistance can be got to the scene quickly either from the front or rear of the incident.

21. The question has been posed about the effect that routing of heavy goods vehicles would have on major traffic and motorway routes. The routing

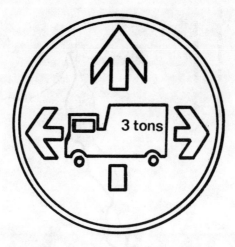

Fig. 3. Lorry routing sign

of HGVs carrying hazardous substances poses a similar question. Is it so desirable to keep certain areas risk-free that other routes should have added to the disadvantages above the further one of the risk of spillage of toxic, corrosive, or inflammable substances? Consider the situation where roadside cafes or lay-bys on the routes selected accommodate at any one time a number of road tankers, each carrying a hazardous substance with a different set of risks. This compounds the hazard, certainly, and what is more, it happens.

22. It may, of course, be possible to control this situation by imposing further constraints on parking of vehicles carrying hazardous substances both on and off streets. But what about the situation when such vehicles are being driven through tunnels or over viaducts or other similar structures? No doubt these features could be taken into account when planning a route, not merely by reason of inaccessibility but because of danger or damage to other users or the structure itself.

23. Experience in Cleveland has shown that not all the factors referred to above can be satisfied fully in all circumstances. The most practical solution has been to devise a coarse routing plan making use of the primary road network (Fig. 4). This caters for the bulk of hazardous substance traffic, leaving route selection between the manufactory, storage place or delivery/pick-up point to the discretion of the vehicle operator, after discussion with the Police and Fire services.

24. This routing plan is linked to a lorry routing scheme being conducted in one part of the county and a preferred route requirement in neighbouring

PAPER 5

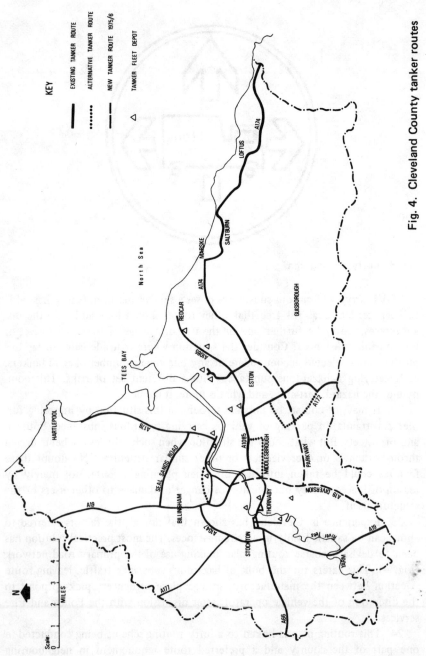

Fig. 4. Cleveland County tanker routes

counties. In this it is interesting to note that areas in which there is no major chemical plant or manufactory may have an equally pressing need for a routing scheme by reason of the volume of hazardous substance traffic passing through. As an example one could look at the mainly rural county of North Yorkshire which lies between petrochemical centres in Cleveland, Humberside and West Yorkshire.

25. The effect of the Cleveland routing scheme, a voluntary scheme to which all the major vehicle operators in the area subscribe, has been substantially to reduce the amount of hazardous substance traffic in town centres and at points of high accident risk. Policing has been found to be minimal and is confined almost entirely to advising drivers from out of the county and who are new to the area. Our judgment on the scheme after some seven years' experience is that it satisfies most of the criteria. We will not willingly seek for the scheme to be mandatory but some such curbs may have to be sought to deal with those who will not comply with the directions to which others voluntarily subscribe.

Uitgave:
Secretarieafdeling Algemene Zaken
Bureau Organisatie Rampenbestrijding
Stadhuis Rotterdam

Fig. 5. Netherlands route sign

PAPER 5

Routing in the Netherlands

26. Constraints on the routes used by drivers of vehicles carrying explosives and dangerous toxic and corrosive substances have been in existence in the city of Rotterdam for some time. This was probably a consequence of the presence near to Europort of a large petrochemical complex and of the volume of traffic between Rotterdam and various other places on the European continent. Operators of vehicles carrying hazardous substances were expected to notify movements, and drivers to conform with the route marking sign. According to the Police Commissioner of Rotterdam the need was recognized, observed, and enforced.

27. A similar scheme has now been introduced by the Netherlands government for use countrywide. The essential elements of the Rotterdam scheme are retained, but the number of substances to which the scheme relates is fewer. Explosives and the most hazardous substances are, however, included and the constraints affect movement of loads in excess of 250 kg. The routes are marked with a distinctive sign (Fig. 5). It has been argued that a comparison with conditions in the UK is not really apt because of differences in size of area and traffic volumes. Nevertheless it is difficult to escape from the thought that what can be done elsewhere can be achieved here—given the will.

Summary

28. I began this Paper by asserting that in the interests of public safety some constraints on the movement of hazardous substances are necessary. The number of reported incidents arising from such movements may well be small, but the volume and nature of the substances moved are sufficient in my view to justify this assertion.

29. The UKHIS provides an effective response to incidents involving spillage of hazardous substances, and there is now a greater awareness of the risks involved. Nevertheless as with all risks it is wiser to pre-empt their happening rather than to deal with their effect.

30. Constraints on the use of selected routes by vehicles carrying bulk loads of hazardous substances involve considerations of the vehicle and its effects, for good or evil, and the movement of a variety of substances, each with a hazard which could be inimical to public safety. Although for many good and valid reasons it may not be opportune to introduce a national lorry routing scheme, this proscription does not necessarily hold good for bulk hazard substance traffic.

31. The problem is how best to devise and arrange route constraints so that the risk to public safety is sensibly balanced against the economic and social needs of the community. Schemes like the voluntary one in Cleveland or the

mandatory Netherlands system point to one form of answer, but these may be considered too parochial or too small in area from which to draw any conclusions which would hold good throughout the UK.

32. Nevertheless, the effects of a large-scale incident stemming from the spillage in volume of a hazardous substance in a public place are potentially so horrendous that no conscience should be stilled by thoughts of the difficulty of finding a workable solution. If, in the interests of public safety, the need for route constraint on heavy goods vehicles carrying hazardous substances is willed, is not willing the means a logical and demanding conclusion?

6. Monitoring, including control systems, regulations, codes of practice and future action

E. J. WILSON, BSc, PhD, FRIC, Head of Dangerous Goods Branch, Department of Transport

Work by the United Nations Committee of Experts on the transport of dangerous goods by all modes is discussed against acceptance by the international organizations, and including additional hazardous materials such as liquid metals and pollutants. The extent to which the UK has developed national regulations, together with work in hand under the Health and Safety at Work etc. Act are covered. Available statistics on the pattern of dangerous goods transport are dealt with with some examples of what is lacking in this field. The Paper includes such problem areas as rail/road priorities, governmental views on routeing, intermodal problems, aspects of safety in construction of vehicles and potential marking systems, leading to thoughts for the future.

Whereas a small number of deaths on the roads attributable to the carriage of petroleum had occurred over previous years with little response from the general public, simply because of a general familiarity with petroleum and its effects, it was the macabre death of a woman in oleum on the M6 on a foggy November day in 1971 which really gave birth to the present concern and fear of incidents on the roads with dangerous goods.

2. It is right and proper that responsible bodies should become familiar with the problems at large, certainly to ascertain what is being done and what needs to be done to allay the fears of the general public and to ensure that transport is reasonably safe—no operation is 100% safe. How 'reasonable' can be defined will emerge to a certain extent as this symposium progresses.

3. It is worth repeating an extract which appears in the Recommendations prepared by the UN Committee of Experts on the Transport of Dangerous Goods: 'The transport of dangerous goods is regulated in order reasonably to prevent such goods from causing either accidents to persons or damage to the means of transport employed or to other goods. Some explosions occurring in the course of transport have indeed ravaged cities. At the same time regulations must be so framed as not to impede the movement of such goods, at least of those which are not too dangerous to be accepted for transport. With this exception, the aim of the regulations is to make transport possible by eliminating risks or reducing them to a minimum. The problem is, therefore, one concerning safety, no less than that of facilitating transport.'

Transport of Hazardous Materials, ICE, London.

4. The title of the Paper was chosen for me. I did not dissent from it, but what does it mean? I suppose in a nutshell, it concerns putting the whole problem in perspective, if this is possible, and providing one person's thoughts of how matters should progress. Such thoughts are my own and not necessarily those of my Department.

5. I believe that the problem principally concerns hazardous chemicals, hence any reference I make to explosives and radioactive substances will be minimal, and I assume that recommendations which will emerge from this symposium will be confined to the chemical classes of dangerous goods.

6. Some of my thoughts will inevitably be repeated in other papers, but such repetition should not detract from the usefulness of the points made.

Dangerous goods

7. To provide a precise definition of dangerous goods in the transport context is not an easy matter. Even such a definition as 'materials which can prove hazardous to man while being transported' would need further elaboration. Does this mean under normal conditions of transport, or does it also mean under accident conditions? Under what types of accident conditions? Would this include the effects of fire, or of being mixed with other materials, hazardous or non-hazardous, or of being mixed with water? Would one expect paper to be classified as dangerous goods? It can easily burn, and under confined conditions can lead to fumes which cause choking; would one expect hay to be a dangerous material? It can heat up and lead to a fire; indeed it is classified in the UN list, and some international regulations regulate its transport. There are such materials as polyurethane foam used in the furniture trade; when this material burns it produces toxic fumes, but would one want furniture so manufactured to be classified?

8. The United Nations Committee of Experts on the transport of dangerous goods has over the years provided a fairly simple classification of dangerous substances into nine major classes with the minimum of precise criteria. The classification appears in Paper 2.

9. So far about 2000 chemicals have been added to the list of dangerous substances, and more are added every 2 years when the experts meet in Geneva to ratify work done by two groups, the Committee of Experts on Explosives and the Group of Rapporteurs.

10. As most chemists are aware, the number of new chemicals devised and produced each year throughout the world is quite appreciable, but the experts would be disinclined to classify all the hazardous ones until these depart from the realms of a laboratory curiosity and are transported in significant quantities. Significant may, of course, mean simply small vials of very hazardous infectious

substances, or it could mean bulk (ship) quantities of fish meal which offer no significant hazard when packed and transported in a few bags, but lead to possible spontaneous heat emission in bulk.

11. Mention was made earlier of unusual substances, e.g. hay, being included in the dangerous goods list. Some of more obvious hazard, such as molten metals like aluminium and steel, are not included. Generally, substances included in UN lists are those likely to be transported across frontiers: molten metals are not in this category. Some are classified for one mode of transport only, e.g. the fish meal mentioned above, and also bulk quantities of iron swarf which can extract oxygen from a sealed hold of a ship and cause asphyxia to anyone entering the hold.

12. In recent years, another class of substance has caused concern when considering the environment as a whole and not just man, namely pollutants. Sea transport was the first mode to be concerned with the effect upon fish when substances enter the sea during incidents during transport or through deliberate cleansing of tanks of bulk chemical ships at sea. Many dangerous goods are pollutants as well as being hazardous to man, but some are pure pollutants. The UN now lists pollutants e.g. polychlorinated biphenyls, UN no. 2315 and placed in Class 9. So far the IMCO Dangerous Goods Committee concerned with packaged goods has declined to consider pollutants as a separate class, nor indeed to give them a separate label. They will remain in the class appropriate to other hazards unless they are pure pollutants, and reference to pollutant effects will appear in the properties part of the schedule.

Statistics

13. Little is known of the general pattern of dangerous goods transported in the UK. The question can be asked, 'Is such information of any benefit to anyone, particularly as one can visualize that dangerous goods are likely to be concentrated in the Midlands and Teesside simply because the chemical industry is concentrated in those areas'.

14. The Department of Transport (DTp) undertook a small exercise in 1973 in order to have some facts which could be used to answer queries being raised at that time. About 500 chemical and petroleum companies were asked to indicate the quantities of chemicals, of particular classes, transported in bulk during a period of 1 month in 1973, and to indicate the origin and the destination of the transport operation. From the exercise, the DTp deduced that about 34 million tons of dangerous goods in bulk are transported on the roads per year, two-thirds of which are petroleum, and the total number of loads per year was estimated to be about 2.5 million. On British Rail the amounts are about 9 million tons/year, of which 6 million tons are petroleum. (These figures should not

be confused with those given in Paper 5 which include all chemicals and petroleum products whether they are hazardous or not.)

15. The busiest roads were those from Merseyside to the M6, with about 100 000 tons/month and 5400 loads/month. Teesside to the A1 was a close second.

16. There is no statutory requirement in UK to report accidents during transport with dangerous goods, except under Section 13(2) of the Petroleum (Consolidation) Act 1928, where reporting is necessary if the accident causes loss of life or personal injury. From the reports which used to be issued annually by Her Majesty's Inspectors of Explosives (the last one being the 99th, issued at the end of 1974), it is seen from the 44 accidents reported in 10 years, that there was a total of 7 deaths. Using the figures quoted in §§ 14–15 and assuming these were the average for the past 10 years, this works out at 1 death/3 000 000 journeys, or 1 death per 3 000 000 000 ton miles, assuming an average journey of about 50 miles/journey. Compare these figures with the 7000 deaths/year from road traffic accidents, and one has to ask oneself what is the justification for more control of dangerous goods vehicles?

17. However, no matter how infrequent serious accidents may be, or even if they are non-existent, as is the case so far with radioactive materials, any incident, no matter how trivial, leads to criticism by the public at large.

18. There have indeed been less spectacular incidents involving spillages etc., which occur almost daily in the UK. Mr Ashton quotes a chemical industry estimate of 200–300 incidents/year. Nearly every one of these leads to delays in traffic—an economic burden to the country, and generally more costly than accidents with non-hazardous goods or motor cars. An incident involving a tanker carrying ethylene oxide on the M5 led to the closure of a section of this motorway for 11 h, and costs in terms of time lost by drivers and the extra mileage of diverted vehicles could have been £100 000–£200 000.

Regulations and Codes of Practice

19. Under the Explosives Act 1875, the Petroleum Act 1928, and the Radioactive Substances Act 1948, a number of regulations concerning transporting dangerous goods by road have been made. They are listed in the bibliography. Also included there are those regulations and codes affecting other modes of transport, the international regulations, and other sources of information including industrial manuals.

20. Those regulations produced under the Petroleum Act were somewhat limited, one reason being that powers were restricted to substances which resembled petroleum. Lawyers had allowed corrosive substances to be so regulated, because some of them could lead to fire, but at the same time they could

find no justification for the control of toxic materials under the Act. Prior to the formation of the Health and Safety Executive (HSE), the Home Office, at that time responsible for the control of dangerous goods by road other than radioactive substances (which were and remain the responsibility of the DTp), had prepared draft regulations to control the construction and operation of tanks and vehicles to be used for the transport of inflammable substances and corrosive substances by road. A code of practice concerned with the transport of toxic materials had also been started.

21. However, with the transfer of the Explosive Inspectorate, first to the Department of Employment and then to the HSE, it was decided that the draft regulations would require modification to be suitable for issue under the Health and Safety at Work etc. (HSW) Act, and in any case it was of greater importance to have general regulations to cover all dangerous goods, particularly the toxic materials, inflammable solids, oxidizing substances, etc., not yet controlled.

22. Pressure from other quarters dictated the necessity to concentrate attention on making the voluntary tanker marking scheme mandatory prior to the issue of the general regulations. At the time of writing, the Consultative Document with the long title of *Hazardous Substances (Conveyance by Road) Tank Labelling Regulations, 1977, and Transport Hazard Information Rules* has already been issued and the HSE is considering comments received.

23. The General Regulations, receiving the attention of working parties of the HSE at this time, will cover all classes of dangerous goods, except explosives and radioactive materials. Much of the earlier work of the Home Office Standing Advisory Committee will appear in the new regulations, notably construction of tanks and vehicles, operation of vehicles, packaging, labelling, declaration, driver training etc. The regulations will probably include those bodies concerned with enforcement: the HSE inspectors, the police and vehicle and traffic examiners of the DTp. So far the last have only been involved in enforcing radioactive regulations, and are equipped with radiation monitors to assist them.

24. Packaging performance tests will be introduced. Such tests will conform with those recommended by the United Nations Committee of Experts, and will be applied to packaging prototypes. A draft guide to performance tests was prepared some time ago by the DTp: ultimately test stations will no doubt be approved by the HSE and used by the appropriate departments responsible for the regulations. Ignoring regulations for radioactive materials, the only international regulations so far to accept performance tests are those of IMCO, and even then there exists a 'grandfather clause' which allows the continued use of sound packagings which, through experience, have been found to be satisfactory. Whether this clause will be adopted in national regulations concerning performance tests remains to be seen.

25. Much has been written in the press concerning 'cowboy carriers' — carriers likely to give a bad name to the industry as a whole. The Secretary of State for Transport has indeed studied this problem, and may consider licensing carriers of dangerous goods. The present O licences for heavy goods vehicles are restricted to 3½ tons and above, but many smaller vehicles carry a few drums of hazardous chemicals which can be a problem if not loaded properly. On the other hand there are many materials which offer a lesser hazard, notably domestic paints, aerosols, bleaches etc. A scheme will need to be produced which will be effective and easily enforced, without the need for expensive operation. Is there evidence to show that, when related to the extent of the traffic, a higher proportion of accidents are caused by cowboy operators, or is the probability of an accident simply proportional to the amount of an operator's traffic?

26. Responsibility for developing regulations and codes under the HSWAct is that of the HSE and Health and Safety Commission (HSC); such draft regulations, following due consultation, are passed over to the appropriate department for making. Secretaries of State can, nevertheless, make regulations themselves under the Act, provided they consult the HSC. The transport regulations to be made and likely to be developed by the HSE are restricted to land transport. They will cover roads first, ports as the next most urgent matter, probably followed by rail and inland waterways. The Act does not extent to ships at sea and aircraft in the air.

27. The Department of Trade will continue to be concerned with regulations and codes for the carriage by sea of dangerous goods. For many years, the DoT's Blue Book was the 'bible' for British seamen. However, in recent years there has been a clamour from the General Council of British Shipping and Industry to have one set of documents to consult when on the high seas. So often in the past it was never too clear to a mariner what differences there were between the British Blue Book and the IMCO Code. With effect from early 1978, the DoT will have in operation what has been called the 'mini-Blue Book', a condensed version which simply includes those controls which differ from IMCO, coupled with the IMCO Code. Ultimately, the British Government will whittle down the mini-Blue Book to a bare minimum; at such time its control will not differ from international control.

28. Air transport is the responsibility of the Civil Aviation Authority, which authorizes carriers overflying British territory under the Air Navigation Order. Such authorizations subject the carrier to working to the provisions of the IATA Restricted Articles Regulations when carrying dangerous goods. The associated inter-governmental body, the International Civil Aviation Organization (ICAO), is now studying the IATA regulations in conjunction with the UN recommendations, with the aim of producing regulations or a code like the IMCO Dangerous

Goods Code, based upon which governments would be expected to produce national regulations. Few codes of practice have been prepared by governments covering the carriage of dangerous goods; those of relevance are listed in the bibliography.

29. One notable example prepared in consultation with government is entitled 'Code of Practice for the Safe Transport of Molten Aluminium by Road in Great Britain', and has been developed by the Association of Light Alloy Refiners. The code includes details such as construction of the ladles, special vehicle features (e.g. fitting of anti jack-knifing devices), inspection procedures, driver training, and notification of emergency services particularly of the Police in the event of special routeing.

30. I have included in the bibliography a number of codes developed by industry, but this by no means covers every relevant code, and apologies are offered in advance if some important ones have been omitted.

31. Reference to regulations and codes of practice is not complete without reference to the international regulations affecting road transport, notably the European Agreement concerning the International Carriage of Dangerous Goods by Road (ADR). ADR, with its Annexes A and B, was first published in 1967, and so far 16 member states including the UK subscribe to this. British regulations differ in certain aspects from the ADR but goods entering the 16 countries are acceptable if they are carried in accordance with ADR, and as long as they also conform with the maritime conditions. The Common Market (EEC) has shown interest in the transport of dangerous substances, and it was proposed that ADR should be mandatory within the EEC, but in meetings held a few years ago, most member states recommended that the time was not ripe for such restricted control.

Emergency action

32. Paper 2 covers emergency problems adequately, but it is worth introducing further information. The British system, THIS, is only really applicable to sea transport when bulk chemicals are carried in road tankers on roll-on/roll-off ferries, and even in those cases some aspects of the emergency action may not be appropriate. IMCO has decided that all schedules for individual hazardous substances will include a section on emergency action; no doubt within the next few years grouping of schedules to cover common emergency actions will be made.

33. The United Nations continue to look for a multi-modal system, as pointed out in Paper 2. At the time of writing the system proposed is very similar to the UK hazard information (HI) plate except that Hazchem is not used, but this action code is left to the discretion of the country concerned, and the hazard

diamond includes the properties code, a compromise between the ADR (Kemler) code and the USA HI code. There is, however, a strong encouragement by the international fire authorities to use the British Hazchem Code on international plates.

34. Unfortunately, despite the arguments put by the UK to have the Hazchem coding made mandatory in the UN system, at the meeting of the UN Rapporteurs in summer 1977 there was little support for us. Indeed, as I understand the situation, the French intend to issue a counter-proposal, to revert to what is already mandatory in ADR, namely to replace the national action code section by the properties code, which was recommended by the UN experts to be in the hazard diamond.

35. Although the transport hazard information system (THIS)—particularly its Hazchem aspect—is appropriate to road and rail transport it is not appropriate to seaports, where there is an interface with IMCO. Some fire authorities in the UK would like to extend Hazchem into the storage areas of ports. No doubt the use of the code on large storage tanks is much akin to having it on a road tanker vehicle, but to use the code where packaged dangerous goods are stored, or in storage compounds containing freight containers enclosing packages of dangerous goods, would be unsuitable. Segregation in accordance with IMCO principles, which is generally applied at ports, is based upon segregation by the colour of the hazard diamond; for instance organic peroxides and oxidizing substances can be stowed together, as can inflammable liquids and inflammable solids, but inflammable liquids must be kept away from corrosive substances, oxidizing substances, substances emitting inflammable gases when wet, etc. Within each of the various classes of dangerous goods are substances with most of the variants of Hazchem, which could only lead to the most restrictive Hazchem coding of all, namely that demanding the use of dry agent and full breathing apparatus, containment and consideration of evacuation, being applied in the port.

36. Following the publication of the voluntary tanker marking scheme and the adoption of it by most of the chemical and petroleum industry, the DTp issued a circular to all highway authorities who might be concerned with the spillages of hazardous chemicals on the highway. The circular informs them of the arrangements for the identification of dangerous goods, of the hazards that may arise, and for the summoning of assistance to deal with an emergency. The CIA's Chemsafe scheme is explained, as is also the associated Road Haulage Association scheme to supply road tankers to transfer liquids from damaged tankers, and to provide suitable equipment for the recovery of damaged vehicles. Mention is made of the co-operation of members of the National Association of Waste Disposal Contractors. In addition, the national arrangements for incidents

involving radioactivity (NAIR) scheme for accidents with radioactive substances is referred to.

Problem areas and the future
Rail/road issues

37. So often during the past 3 or 4 years letters to MPs and Parliamentary Questions (PQs) have included the statement that dangerous goods should be transferred to the railways as the safer mode of transport. There has been much debate within governmental circles, including the HSE, on this issue, and in general governmental policy remains that there is really no justification for such a transfer. There may well appear to be fewer accidents with dangerous goods on the railways in the UK, although the number of serious accidents by road is fairly small when compared with the amount of traffic. Catastrophic accidents could occur on rail, where large amounts of the same type of product may be carried at the same time, and particularly where the traffic could be travelling at fairly high speeds through urban areas.

38. The problem has been debated at RID/ADR meetings recently when one or two continental countries wished to forbid the transport of large quantities of chlorine by road, but allow it by rail. The chemical industry in the UK prepared a paper, which was supported and presented by the government, supplying statistical evidence to substantiate a claim that the hazard was much the same no matter what the mode of inland transport. The joint meetings of RID/ADR turned down the proposal to limit the mode of transport for chlorine, and it was learned later that the country which made the proposal was simply anxious to control the frontier traffic of the gas. If it was not allowed in the international regulations, then bilateral agreements would have to be made, and in this fashion the government would know the extent of the traffic and, although not forbidding transport by road, would have more control over movement of the gas. There may be some merit in the adoption of such a procedure, but it is entirely against the spirit of the conventions, and to date has no justification on safety grounds.

39. Despite the arguments mentioned above, there would be additional problems should the government adopt a policy of rail in preference to road. The majority of chemical companies do not possess rail heads nor indeed do their customers. To transfer hazardous liquids from a tanker vehicle to a rail tanker away from factory premises is not without its hazards, and generally speaking it is preferable on safety grounds to have only one loading and one unloading operation per consignment. With the greater use of tank containers, i.e. portable tanks which can be held within a framework which acts as a freight container and hence can have multi-modal use, the transfer problem can be alleviated, thus

allowing transport by road, rail, sea, etc. during the movement of one consignment.

Routeing

40. Paper 5 gives views on routeing. This is obviously one area for further study, and it is worth adding a few words to what Mr Ashton has said. Government has adopted a fairly firm line on this subject in recent years, but this does not mean that such a line is sacrosanct: public opinion can very quickly change a governmental view! White Papers have appeared on the subject of routeing all types of heavy lorry, and generally government has considered that dangerous goods lorries do not present a special case. Paper 5 covers the general arguments, and far be it for me to elaborate upon them, other than to say government will want further justification before suggesting that dangerous goods vehicles are special.

41. This brings me back to more statistics, evidence in fact that there is likely to be a decrease in the incidents in congested areas of towns referred to in Paper 5. Are there really significant incidents of spillage in congested areas, where vehicles are likely to be 'crawling' along anyhow. There have even been 'demands' from the public that dangerous goods should be taken off the motorways on to minor roads, when the odd bad accident occurs. Government has included in answers to PQs the fact that industry is encouraged to contact local Police and Police in the destination area for advice as to recommended routes for dangerous goods lorries at particular times of the day. This voluntary scheme is very effective in Cleveland but how effective it is in other areas is not known.

42. Hazards of dangerous goods in tunnels is another problem area. Some of the larger tunnels under rivers in the UK are toll tunnels, and by-laws as to their use are the responsibility of the local authority. The DTp approves such by-laws, but has no power to prevent a local authority from adopting by-laws more rigorous than any transport regulations, if it so wishes. Some of the by-laws appear to be unduly restrictive, notably in respect of the banning of certain radioactive traffic, and yet half a dozen tanker vehicles of petroleum spirit in one convoy, under escort, will often be seen during the day when the tunnel appears to be congested. It is also possible for the local authority not to consider the broader issue: that to prevent a particular load of dangerous goods from proceeding by the quickest route to its destination may in certain cases lead to a higher probability of the goods being involved in a serious accident, because of the extra length of journey, perhaps via a bridge and through congested urban districts. This is an area for further study. Naturally the tunnel authorities have to be ultracautious; a tunnel could be severely damaged, which could lead to severe loss of revenue, or to severe congestion elsewhere if a serious accident blocked the tunnel.

Vehicle construction

43. How far should one go into the safer construction of vehicles? A tanker vehicle carrying inflammable liquids, overturned at speed, can be penetrated by curb stones or other sharp objects, and will inevitably burst into flames. The oleum tanker referred to earlier, despite being staunchly designed, was penetrated by the sharp corner of another vehicle. Should tankers have more protection to prevent this penetration? Should the use of aluminium tanks for inflammable liquids be discouraged because a few incidents have led to the tank melting completely, the melting point of aluminium being below that of burning hydrocarbons? Should the use of rigid vehicles rather than articulated vehicles be encouraged, because there appears to be a tendency for articulated tanker vehicles to overturn a little more frequently than rigid vehicles? Should the chassis of tanker vehicles be equipped with standardized means to allow the emergency services to bring them upright more safely after accidents?

44. Should there be more control over the filling of tanker vehicles, since some incidents, particularly in the USA, have indicated that liquid surge in tanks, say, only three-quarter filled, has contributed significantly to the overturning of vehicles? Should the use of glass reinforced plastic (GRP) tanks for inflammable liquids be encouraged? Should tanker vehicles have their tanks filled with a mesh of thin expanded aluminium foil, as has apparently been successfully applied to drums, in order to reduce the effects of explosions in serious fires; should the advantages of double skins as applied to certain chemical ships be examined or would this be too costly and not justified on the grounds of probability of serious accident? Would double-skinning present other danger problems because of the additional weight, or the difficulties of inspection and maintenance?

45. The ADR regulations call for the provision of a master switch in a dangerous goods vehicle which will allow all the electrical circuits except the tachograph to be isolated from the battery, this master switch to be accessible both inside and outside the vehicle. It is highly probable in an incident involving spillage of inflammable liquids that the master switch would be operated, but this, of course, leaves the vehicle very vulnerable if stationary on a badly lit road at night. The ADR regulations overcome this problem by insisting that dangerous goods vehicles should also carry amber lights for use in such circumstances; but UK legislation forbids this. Only the emergency services can use such lights in accident circumstances. The UK has argued in the past against the provision of amber lights because with their infrequent use probably the battery would be flat on the odd occasion when the light might be needed. Is this thinking still applicable?

Other considerations

46. Regulations concerned with the transport of packaged dangerous goods by road vehicle do not generally include any provisions relating to loading and securing the load. It has always been considered that codes issued by various bodies for general road loads are as applicable to dangerous goods as to non-dangerous goods. Incidents of hazardous drums falling off lorries hit the head-lines, but do the number of such incidents warrant special treatment for dangerous loads?

47. Mention has been made of the drainage systems of roads carrying a large volume of dangerous goods traffic. Water authorities are worried by the number of incidents on the roads, leading to the pollution of water systems. Despite the fact that the THIS system provides guidance to emergency authorities as to which substances need to be contained by, say, sandbags on the highway, so preventing entrance to drain systems, significant quantities of pollutants could reach water systems prior to the arrival of the emergency services. With many drainage systems water authorities can take speedy action to prevent pollution, provided they are informed early enough, but there may be vulnerable systems where immediate action may not be speedy enough, and the provision of appropriate delay traps may become necessary.

48. Is there a need for a driver of a dangerous goods vehicle always to have an assistant travelling with him, so that at least one of them would be capable of warning others, following an accident? Such was the requirement in ADR for certain of the more hazardous dangerous goods, but this has been relaxed recently, and only applies in respect of certain explosives should the competent authority of the country(ies) concerned require it.

49. Should drivers of dangerous goods vehicles be of a mature age, and must they also have a special medical examination, repeated annually? Few countries favour such ideas; far more important is it for the driver to be trained adequately in the hazardous aspects of his load. A fit man can have a coronary without a medical inspection spotting the problem in time.

50. Is there a need for vehicles carrying dangerous goods to be more readily identified for safety reasons than with the present hazard diamonds or HI plates? A spate of fog incidents will inevitably lead to MPs being encouraged to ask for special flashing lights for identification purposes, forgetting that a flashing light in dense fog could attract members of the public who might mistake it for the emergency services. In any case, if the dangerous goods vehicle was involved in a severe accident, capable of causing a spillage, no doubt the flashing light would fail to function. Suggestions have been made that dangerous goods vehicles should all be painted with a bright and distinctive colour as a warning to the

general public. Such an idea may appear attractive in theory, since only 5% of goods carried on the road are hazardous (see Paper 5). However, since many vehicles can carry non-hazardous goods one day and hazardous goods the next the exercise would lead to virtually all vehicles being distinctively coloured.

51. Finally, a discussion on the transport of dangerous goods is not complete without some reference to publicity and training. The HSE and the DTp have the question of publicity in hand, and are considering the suitability of films, the use of filler films on TV, the use of posters (particularly in motorway service areas), the use of local radio, inclusion of the hazard diamonds and the HI plate in the Highway Code, etc. As for the training of transport personnel, much is being done by industry and the Industrial Training Board for road drivers, but more could be done to encourage training in the handling of dangerous goods within ports, both sea and air. How much training is indeed encouraged and organized by the TUC itself?

Summary

52. Where are the principal areas for further study in the transport of dangerous goods field?

53. There is a need for a more comprehensive study of the pattern of dangerous goods transport than that undertaken by the Department of Transport, even if confined to the 'chemical' areas only, combined with a study of special routeing.

54. Little if anything can be learned from the serious accidents involving dangerous goods, but if there are 300 minor incidents/year, some pattern should emerge. Many of the main chemical transporters use designs of vehicles complying with the new regulations to be enforced by ADR, and in keeping with the draft regulations prepared by the old Home Office Standing Advisory Committee. No doubt investigations could ascertain the adequacy of these new regulations from a study of the incidents. In the absence as yet of the statutory need to report incidents, some voluntary arrangement would need to be arranged.

55. Incidents of pollution of waterways from various sources, including transport, are collated by the Department of the Environment. A study in conjunction with the water authorities could possibly lead to the ascertaining of those highway drains which are vulnerable in the event of chemical spillages, and which perhaps require delay traps.

56. Will the forthcoming protection arrangements for tank vehicles in the event of overturning (new marginal 211 127 of ADR) be adequate enough? Is sufficient research being undertaken in this field, and in other fields of tanker vehicle construction from the safety point of view?

57. How adequate and how standardized is driver training, and should this be extended to other persons handling dangerous goods in sea and air ports?

58. Finally is more publicity needed to alert the general public?

Bibliography: Regulations and codes concerning conveyance

UK Regulations: road

Regulations made under the Explosives Act 1875

Order of Secretary of State No 11, dated 20 September 1924, making bylaws as to conveyance of explosives by road.

Order of Secretary of State No 11, dated 11 December 1939, relating to the conveyance of detonators and electric detonators with other explosives.

The Packing of Explosives for Conveyance Rules, 1949 (SI No 798).

The Packing of Explosives for Conveyance Rules, 1951 (SI No 868).

The Conveyance of Explosive Byelaws, 1951 (SI No 869).

The Conveyance of Explosives Byelaws, 1958 (SI No 230).

Regulations made under the Petroleum (Consolidation) Act 1928

(a) *Petroleum spirit and petroleum mixtures*

The Petroleum (Mixtures) Order, 1928 (SRO No 993).

The Petroleum Spirit (Conveyance by Road) Regulations, 1957 (SI No 191).

The Petroleum Spirit (Conveyance by Road) Regulations, 1958 (SI No 962).

The Petroleum Spirit (Conveyance by Road) (Amendment) Regulations, 1966 (SI No 1190).

(b) *Carbon disulphide*

The Petroleum (Carbon Disulphide) Order, 1958 (SI No 257).

The Carbon Disulphide (Conveyance by Road) Regulations, 1958 (SI No 313).

The Carbon Disulphide (Conveyance by Road) Regulations, 1962 (SI No 2527).

The Petroleum (Carbon Disulphide) Order, 1968 (SI No 571).

(c) *Compressed gases*

The Petroleum (Compressed Gases) Order, 1930 (SI No 34).

The Gas Cylinders (Conveyance) Regulations, 1931 (SI No 679).

The Gas Cylinders (Conveyance) Regulations, 1947 (SI No 1594).

The Gas Cylinders (Conveyance) Regulations, 1959 (SI No 1919).

(d) *Other substances*

The Petroleum (Inflammable Liquids) Order, 1971 (SI No 1040).

The Inflammable Liquids (Conveyance by Road) Regulations, 1971 (SI No 1061).

The Inflammable Substances (Conveyance by Road) (Labelling) Regulations, 1971 (SI No 1062).
The Petroleum (Corrosive Substances) Order, 1970 (SI No 1945).
The Corrosive Substances (Conveyance by Road) Regulations, 1971 (SI No 618).
The Petroleum (Organic Peroxides) Order, 1973.
The Organic Peroxides (Conveyance by Road) Regulations, 1973 (SI No 3221).

Codes

Home Office Code of Practice for the Packaging of Dangerous Substances for Carriage by Road.

Regulations made under the Radioactive Substances Act 1948

The Radioactive Substances (Carriage by Road) (Great Britain) Regulations, 1974 (SI No 1735) (replaced regulation 1970 SI No 1826).
The Radioactive Substances (Road Transport Workers) (Great Britain) Regulations, 1970 (SI No 189).
The Radioactive Substances (Road Transport Workers) (Great Britain) (Amendment) Regulations, 1975 (SI No 1522).
Code of Practice for the Carriage of Radioactive Materials by Road, 1975 (ISBN 0 11 550348).
Code of Practice for the Storage of Radioactive Materials in Transit, 1975 (ISBN 0 11 550366 8).
Code of Practice for the Carriage of Radioactive Materials through Ports, 1975 (ISBN 0 11 550371 4).

Regulations made under the Health and Safety at Work, etc. Act 1974

Draft Consultative Document Hazardous Substances (Conveyance by Road) Tank Labelling Regulations, 1977, and Transport Hazard Information Rules.

UK regulations: other modes

Rail

British Rail List of Dangerous Goods (LDG) and Conditions of Acceptance by Freight Train and by Passenger Train (BR 22426, 1977 Edition).

Sea

The Merchant Shipping (Dangerous Goods) Rules 1965 (SI No 1067, as amended by SI No 332, 1968 and SI No 666, 1972).
The Report of the Standing Advisory Committee on the Carriage of Dangerous Goods in Ships 1966: the 'Blue Book'. 2nd edn, 1971.

Air

The Air Navigation Order 1974 (SI No 1114).

PAPER 6

International regulations and codes

UNITED NATIONS. Transport of Dangerous Goods. Recommendations prepared by the Committee of Experts on the Transport of Dangerous Goods (ST/SG/AC10/1). United Nations, New York, 1976.

Intergovernmental Maritime Consultative Organisation (IMCO) Dangerous Goods Code.

IATA. Restricted Articles Regulations published annually. International Air Transport Association, Geneva, published annually.

European Agreement concerning the International Carriage of Dangerous Goods by Road (ADR), and Annexes A and B. HMSO, 1976.

International Convention concerning the Carriage of Goods by Rail 1961 (CIM). Annex 1—Regulations concerning the Carriage of Dangerous Goods by Rail (RID). HMSO, 1977.

Some additional codes and manuals

Report of Her Majesty's Inspectors of Explosives (99th issued 1974). HMSO.

CIA. Code of Practice for the Safe Carriage of Dangerous Substances in Freight Containers. Chemical Industries Association Ltd.

CIA. Road Transport of Hazardous Chemicals, a Manual of Principal Safety Requirements. Chemical Industries Association Ltd.

CIA. Transport Emergency Cards. Chemical Industries Association Ltd.

ALAR. Code of Practice for the Safe Transport of Molten Aluminium by Road in Great Britain. Association of Light Alloy Refiners.

CIA. Chemsafe: a manual of the Chemical Industry Scheme for Assistance in Freight Emergencies. Chemical Industries Association Ltd.

CIA. Hazard identification: a voluntary scheme for the marking of tank vehicles conveying dangerous substances. Chemical Industries Association Ltd.

CIA. Hazchem codings: allocated by the Joint Committee on Fire Brigade Operations and Confirmed by the HSE. Chemical Industries Association Ltd.

CIA. UN Numbers (Alphabetical and Numerical). Chemical Industries Association Ltd.

PLA. Schedule of Dangerous Goods. Port of London Authority.

TEES & HARTLEPOOL PORT AUTHORITY. Petroleum (Consolidation) Act 1928, The Tees & Hartlepool Harbour (Petroleum) Byelaws 1976. Most ports have their own byelaws controlling dangerous goods.

NAIR (National Arrangements for Incidents involving Radioactive Materials)—Handbook. National Radiological Protection Board (NRPB).

Department of Transport. Distribution pattern of dangerous goods transported in bulk by road (WP/TDG(74)39).

Department of Transport. Spillage of Hazardous Chemicals on the Highway—Circular ROADS 21/76.

RHA. Bulk liquids functional group—emergency load transfer scheme. Road Haulage Association Ltd.

Department of the Environment. Code of Practice: safety of loads on vehicles. DoE, London 1972.

Discussion: Papers 5 and 6

The Chairman (Dr W. E. Duckworth)
Discussion on the earlier papers should have generated the sense of unease which the organizers had planned, leading to consideration of the means available to reduce the risk of major incidents. The problems are real, the potential hazards are numerous, and the interest by the informed and by the lay public is increasing. After the previous discussion, I shall be much more aware, when I am overtaking some large vehicles on the motorway, of the meaning of the Hazchem code, and that will certainly condition my speed and space of overtaking.

2. The responsibility for improving the present position rests not only with the legislators in the UK and abroad but with the public services for implementing the law and mitigating the effects of eventual accidents, and especially with the professional scientists and engineers who must assess the risks and provide data in a reasonably usable form for those who do not have the technical knowledge on which judgment and action can be based.

3. One of the primary objectives of the discussion is to summarize the action which has already been taken and to underline the need for continuing collaboration between the different disciplines concerned. It will also, I think, emphasize human frailty which has already been discussed and of which I have had experience in my own organization. We once investigated a situation at a small electronics factory where the ladies in the assembly shop were complaining of nausea by mid-afternoon. Investigators detected the presence of trichloethylene. In an adjoining part of the factory a trichloethylene degreasing plant had recently been installed. The extraction system was superb, but in tracing the ducting it was found that this ended 6 in. short of the roof.

Mr J. Farrall (Chief Environmental Health Officer, Cheshire County Planning Department)
It is most important to realize that the chemical industry is one of the UK's major growth industries and has the second largest output of any industry in this country. As with all problems, prevention is much better than cure. With proper planning, the number of hazardous loads requiring transportation could be reduced, particularly if the downstream processes requiring these facilities for the disposal of hazardous waste were more strategically located. There is little chance of dealing with existing premises, but it would be possible to control future development using statutory planning controls. The vital factor of transportation should be fully considered when structure plan policies are being evolved by county councils, and adequate strategically located land should be allocated to cover the future needs of the chemical and petrochemical industry.

DISCUSSION: PAPERS 5 AND 6

This would mean that pipelines could be introduced to supersede the carriage of materials by road.

5. The problem of shopping areas and road tunnels is mentioned in Paper 5, but to these two features I would like to add bridges across major rivers. In the NW, the Runcorn-Widnes bridge and the motor bridges at Warrington cause road traffic containing a high proportion of hazardous loads, as indicated in Paper 6, to converge on strategically placed towns. Mr Marsden earlier amplified the problems that local authorities see regarding the conveyance of hazardous loads through urban areas. During the summer of 1975 the Cheshire Councils commissioned a survey of the Cheshire area. The subsequent report[1] included the environmental aspects of the chemical and allied industries, including transportation. It was the section dealing with road transport which caused most comment. It was not at all difficult for the local press to print photographs of tankers conveying hazardous loads through high density urban areas in support of the various articles that appeared on the publication of the report.

6. In Cheshire some routing of hazardous loads has been arranged by the more enlightened companies, but even they admit that they have problems with their haulage contractors, even though the routes are written into the contracts. It would appear that routeing should be made obligatory. The suggestion that dangerous goods lorries do not present a special case is completely unacceptable to the local authorities. To compare a milk tanker with a similar vehicle conveying hazardous materials is not facing-up to the real situation. It was noted during the survey that drivers on piece work had a tendency to fast and sometimes reckless driving. In those circumstances, the suggestion that carriers transporting dangerous substances should be licensed has much to commend it. One company has the philosophy that it is safer for vehicles to travel at slow speed through town centres rather than at high speed on a ring road.

7. The hazard of unattended overnight parking in town centres and residential areas was also highlighted as one of the particular problems causing major concern to the county fire officer. One factor is that drivers travelling on motorways make use of the overnight sleeping facilities at adjacent towns. This does not appear to be a problem in Cleveland, but it certainly is a problem, particularly in the towns adjacent to the M6 motorway.

8. Concern was also expressed in the report at the potential danger of polluting potable water supply. There is a particular problem in Cheshire as the River Dee is a salmon river, and provides most of the water supply for Merseyside and Chester. If this potable water, which is controlled by the Welsh National Water Authority, should become polluted by a load of hazardous material, there could be a major problem.

9. Our experience indicates that the voluntary codes should become man-

datory. Also, in order that an appraisal of the risk situation can be made, detailed records of road accidents involving hazardous loads should be kept. This is now being done in Cheshire, where there are about 6-10 accidents of this nature each year.

10. The report also gave detailed information on the movement of hazardous materials associated with the chemical industry. It also locates the main routes and thus identifies the vulnerable areas within the county. This inventory is broken down into the eight Cheshire districts. It is significant that four of the 19 recommendations involve road transport.

Mr S. T. Jones (Consultant)
For 10 years I was the Chief Executive of the Mersey Tunnels. Clearly Dr Wilson is right (Paper 6, §42) to call attention to the tunnel situation, which is unsatisfactory. Tunnels in the UK have different by-laws controlling dangerous goods, different practices, and, of course, vastly different levels of enforcement. The Mersey, Tyne and Dartford Tunnels have produced a unified set of regulations, but these tunnels are tolled, and therefore manned, and enforcement is easier than on other tunnels.

12. Dr Wilson does less than justice to the local authorities. He comments that tunnels can have their own by-laws, but unless these are approved by the Secretary of State they have no validity and cannot be enforced. They are advertised; there is opportunity for objections to be heard; the Minister may, indeed, order a public enquiry. It is unlikely that the Secretary of State would sign a set of by-laws if at the time they were believed to be wrong or unreasonable. The differences in by-laws in the different tunnels are probably because they were introduced at different periods when the ideas on dangerous goods problems varied.

13. Dr Wilson suggests that the local authorities are ultra-cautious. I would not disagree with that, but to say that a tunnel could be severely damaged and that this could lead to loss of revenue or to severe congestion elsewhere is only a small part of the reasons. Tunnels are different from ordinary highways, in that if there is smoke or gas in a confined space, users in the tunnel have no ready escape. Tunnels in the UK are generally ventilated longitudinally; the polluted atmosphere in the tunnel is moved along the traffic space before removal to a high altitude. Smoke, flame and toxic gases will also move along this space, probably over people who in an incident would be trapped in their cars. If the tunnel operator shuts off the ventilation, the smoke and gas remain in the tunnel, if he turns up the ventilation to blow the fumes away, he could fan the fire, or blow flames and fumes over queueing traffic on either side of the incident. So there is an agonizing decision which has to be taken instantly as to whether to blow the fumes towards the approaching emergency services or

towards traffic trapped at the incident, or to turn off the ventilation altogether.

14. What advice or instruction should be given to the man on the spot, who may be of no more than medium-grade staff? It is this decision which leads tunnel managements to do everything possible to minimize the possibility of a major incident, and is the real motive behind the restrictions on dangerous goods in tunnels. In a tunnel, even inert gas can be dangerous. In the past there has been little guidance from government departments to tunnel operators. There have been differences of opinion between the Department of Transport and the Home Office, and in many cases government departments are reluctant to be seen to be influencing decisions of this nature. If the government departments are able to give written positive advice, the tunnel authorities will be only too pleased to receive it and take notice.

Mr R. H. Lewin (Fulmer Research Institute)

I would like to give three examples of recent work carried out by the Corrosion and Technical Services Department at Fulmer involving the transport of hazardous materials.

16. The first problem related to the transportation of liquid ammonia. As a result of incidents with tankers carrying this material a multiclient project was undertaken to study the possible failure modes of storage and transport containers. The problem is complex. Both mechanical and chemical factors could contribute: the presence of oxygen and water in the liquid ammonia, the type of container material and the range of mechanical deformation envisaged.

17. The point I wish to make is that the chemistry of the container and contents must be studied in association with the mechanical loadings. It is not sufficient to dip a piece of carbon steel into liquid ammonia to determine if they are compatible in the context of the working conditions of a road tanker.

18. The second example related to the failure of a transport vehicle containing sulphuric acid; the investigation showed that deterioration in the tank resulted from the effect of previous loads.

19. The last problem related to the failure in the aluminium alloy shell of an aircraft due to a spillage of gallium metal. While transportation by air is not covered here, until this catastrophic failure caused by a seemingly harmless liquid metal, no one would have objected to transporting this metal in an aluminium alloy container, whether by air or by road.

20. These examples are typical of past and present experiences in the transportation sector, and demonstrate the need for competent professional advice in preventing such incidents. A relatively small investment in such advice could prevent many incidents and reduce the enormous costs associated with the 'down time' during the subsequent enquiry.

DISCUSSION: PAPERS 5 AND 6

21. The questions then that I would ask Dr Wilson in the light of continuing experiences are:

(a) where does a transportation company, not having the technical back-up of the larger haulage groups, turn for information regarding the hazardous nature of loads?
(b) in relation to a particular chemical and container material, who considers the possible failure modes in a tanker?
(c) who checks on the previous history or any remaining residues in a tanker before recharging?
(d) is any audit made of container accidents in which death does not ensue?
(e) in relation to these questions, and assuming all these matters are considered, does the information get incorporated into codes of practice?

Mr C. Bainbridge (Teesside Transport Training Association Ltd)
Little has been said about the standardization of driver training.

23. In the past few months recommendations have been published by the Chemical Industries Association, the Road Haulage Association and the Road Transport Industry Training Board. This group has come up with recommendations on the selection and recruitment of drivers, the operation of the tank itself and its ancillary equipment, product knowledge, legislation, labelling and marking, normal procedures, first aid, causes of vehicle fires, use of fire extinguishers and emergency procedures. There are now five nationally approved centres, one in Cleveland, one in Runcorn and one in Essex. The other two, one in South Yorkshire and one in Chippenham, begin in January.

Mr G. M. Lilly (John Forman Ltd)
I should like to make a brief comment on behalf of the bulk liquid hauliers. Delegates may have got the impression that the industry has done little to regulate itself so far as the carriage of hazardous chemicals is concerned. I would like to pay tribute to the work of the Chemical Industries Association, who have spent time and money in devising schemes, codes of practice and procedures designed to reduce hazards in the carriage of dangerous chemicals.

25. Hauliers too, have done their bit on selection and training of drivers and on introducing codes of practice for the testing and maintenance of vehicles, tanks, hoses and valves. This is most important because the greatest hazards arise when the load is being transferred from storage into tank or being delivered from transport tank into receiver's premises. Many chemical manufacturers and chemical carriers have spent money in building new and safer vehicles and have antici-

119

pated to a very large extent the expected regulatory rules for the construction of vehicles to carry hazardous chemicals.

26. These recommendations and proposals are already in use. They are being monitored and audited, both by the chemical manufacturers themselves and by hauliers. Do not let us re-open subjects which have already been properly dealt with. The main benefit arising from discussion is to pinpoint those areas where the industry has not yet acted adequately.

Dr R. J. M. Willcox (Education Officer, Institution of Metallurgists)
The education of future engineers is of particular relevance here. When I trained as a metallurgist there was little reference to the interaction between hazardous materials and engineering materials, metal or otherwise. One of the things the professional bodies could well do is to investigate the extent to which courses deal with transportation hazards and design problems related to materials used.

28. There is discussion in academic circles as to whether a 4-year degree course is desirable, particularly in the engineering field. One of the views coming forward, certainly in metallurgy and materials, is that the pressures on the current 3-year degree course, particularly from the schools end of the system, are sufficiently large to ensure that very little additional engineering material will be included in a 4-year degree course. Professional bodies could be of assistance in ensuring that in their continuing education programmes courses are put on which deal with this particular subject.

29. Mr George said that manufacturers regard it as their job to ensure proper design and safety and pleaded that there should not be further legislation as the manufacturers could look after this. From the professional body point of view I would ask what steps do they believe they have taken to ensure public confidence in their ability to do this? What seal of competence do they look for in their professional personnel? A means of inspiring public confidence would be to appoint personnel who are members of their professional bodies. One of the terms of reference of the Finniston Committee deals specifically with the possibility of statutory registration, and it seems to me that this particular area is ripe for consideration.

30. This symposium demonstrates co-operation on an interdisciplinary basis between science and engineering institutes. This should continue, because it is particularly important in this field that scientists should be able to communicate with engineers, and communication must be developed within individual disciplines.

Dr D. Train (Cremer and Warner)
This symposium could have been held 3 years ago had it not been for the resist-

ance from the chemical industries. There was a rebuff both from the chemical industries and from the government. The purpose of the meeting is to bring together professional people who have a commitment in their professional charters to devote their services to the interests of the community. It was to demonstrate that the professional engineers and scientists could bring together the variety of scientific and technological disciplines in order to discuss a subject of public concern.

32. We are searching out the present state of the problems, to show that we are not complacent about them and to demonstrate that we would look at these matters objectively. The main intention is to pick up those points which have a policy element in them and to take these back to those institutions which have a particular expertise, in order to examine them in greater depth.

33. For example, for some years there has been a standing technical committee on synthetic detergents. No legislation is involved, but it combines professional institutions and the government in this particular area. Codes of practice will be useful, but do not let us necessarily keep seeking a further regulatory mandatory control.

Mr P. N. Anderson (Imperial Chemical Industries Ltd)
As far as the chemical industry is concerned there has never been any intention to exclude professional engineers from such forums. The publicity for them has been wide and extensive and representatives would have been welcome. The chemical industry, of course, employs many professional engineers, including metallurgists and chemical engineers.

35. I would like to make some brief comments. A milk tanker can be compared with a chemical tanker. The chance of someone being killed by the one rather than the other is only 0.5% less, because the main danger comes from their speed and weight, and there is nothing very much which can be done about that. The second point is that anyone wanting information on ammonia should go to the manufacturer of ammonia.

36. Referring to Mr McMillan's remarks, the chemical industry has become acutely aware of the hazards involved in righting road tankers. Particularly in the case of a pressure barrel which has sustained some damage in turning over, the most careful consideration is needed before moving it or attempting to transfer the load. Local recovery firms should not be used to get it back on its wheels.

Mr J. L. Shelbourn (County Surveyor, Staffordshire County Council)
I assume that since Mr Ashton's scheme was entirely voluntary, he did not have to engage in public participation exercises. If he did, I wonder what were the views of those selected for the privilege of living alongside the routes to be used

for the movement of hazardous materials. The other question I would like to ask is what numbers or proportion of hazardous movements were diverted from unsuitable routes to those selected?

38. I am chairman of a working group of the County Surveyors' Society and was involved with the Dykes Act, a cause which we espoused with great enthusiasm initially. But the stage was reached when it was quite evident that in order to have a satisfactory network for the movement of heavy vehicles, some new roads were needed. There is already a national grid for high loads and one for abnormal heavy loads. It may be possible to devise a grid for hazardous loads. The County Surveyors' Society will be happy to assist in this. However, having been somewhat disillusioned by experience with the Dykes Act, we do not underestimate the difficulties of trying to devise a national or national-plus-county network.

Mr D. Barber (Liquified Petroleum Gas Industry Technical Association (UK))
I would like to make some comments on Mr Lewin's questions, in particular that concerning any audit of accidents without death and subsequent incorporation of the lessons into codes of practice.

40. The Liquified Petroleum Gas Industry Technical Association (UK) (LPGITA) has for a number of years operated a voluntary reporting scheme whereby all incidents involving liquefied petroleum gas (LPG) are recorded and the results analysed to establish any trends which could indicate the need for consumer education or the publication of literature or codes of practice.

41. Two years ago, following a series of accidents including one fatality, resulting from failure of LPG transfer hoses, the Safety Committee of LPGITA recommended that a code of practice should be produced for such equipment. Since that date a draft code has been produced by a working party in co-operation with the hose manufacturing industry. This is being discussed with the Health and Safety Executive prior to publication. Work is also to continue on producing a further code covering the in-service maintenance and testing of such hoses.

42. Following an incident involving a butane road tanker which occurred in January 1974, the Home Office (Fire Department) decided to rewrite their instructions to fire services covering LPG incidents. A working party on which I represented the LPG industry, met during 1975–76 and has produced a document[2] which was issued to all fire services early in 1977.

43. During the deliberations of the working party it was realized that the injection of water into a leaking LPG tanker could in certain cases provide an effective seal to a hazardous leak. It was further realized that such an operation could in some circumstances itself result in further problems and should there-

fore be carried out only under specialist supervision. Clearly further information for guidance was necessary before the emergency services could take even limited action without the specialist assistance being present at the actual site of the incident.

44. In 1976 Shell UK Ltd carried out limited trials aimed at sealing a liquid phase LPG leakage with water which, while successful, indicated that further work was required. Following these trials discussions were held between representatives of the LPGITA(UK), the Institute of Petroleum and the Fire Service Technical College which resulted in a series of very successful tests being carried out at the college during September 1977. This work is in draft but will be published.

45. A booklet on handling emergencies for LPG is in preparation and is to be published shortly.[3] The object is to provide guidance for the handling of LPG emergencies by bringing together the valuable information already available from many independent sources in addition to such information as that given above. All of the work has been carried out, not as a result of legislation demand, but because the LPG industry recognized the need.

46. The LPGITA(UK) may be a little known association but nevertheless it has as its aims the safe development, manufacture, application, storage and transportation of liquefied petroleum gas, and publishes the results of all such study and research.

Mr Ashton (Paper 5)
Mr Shelbourne made a very valid point because participation, in the sense in which he meant it, is currently in vogue. Fortunately when Cleveland started the scheme seven years ago, people had never heard of it. Or if they had, certainly the chairman of the group which was then leading the scheme, had never heard of it. The group did undertake a form of participation, in that the people we selected were from a wide variety of backgrounds, including trade union drivers and hauliers and anyone who might have a part to play in this. We brought the local newspaper into it and let them publish what we had in mind in the first place.

49. As Mr Shelbourne quite rightly said, those who had been selected for the routes were not very kind in some of the things they said about us. However, I think, only one route had to be amended in order to meet this pressure, because when we looked at the route we realized that we had made a mistake. We published the map; the greatest possible local publicity and we received only praise, because it was produced by the people who were interested in the subject, and who were actually working on it and knew what it was all about.

50. Mr Cumberland suggested that the scheme had not begun in 1975, but in fact it has been running since 1974.

51. There was not a great deal of diversions of hazardous movements, except in the three centres of population, where it was fairly considerable. In the seven years since the scheme started, I can honestly say that I do not know of one instance where we have seriously had to take to task any haulier or manufacturer about taking any of these substances through one of those centres. I will not pretend that there have not been breaches, there have but we have been able to deal with them by direct reprimand. By having the co-operation of the people in the area, the scheme policed itself to a large extent. To that extent, I think there was public participation.

Dr Wilson (Paper 6)

The purpose of the Paper was to provoke discussion. It really is a series of questions and not a set of recommendations.

53. In respect of §42, I am happy that Mr Jones made his remarks about tunnels. So far as concerns the making of by-laws for tunnels, as I write in the Paper, as long as they are equivalent to, or more stringent than road regulations, the Department of Transport has no qualms about approving such by-laws.

54. Mr Lewin's questions will be answered by Dr Jeacocke, who is a chairman of the Health and Safety Executive Working Party dealing with the preparation of regulations. This has done all the work on the tanker marking regulations and is now working on the general regulations which will come out during 1978.

Dr G J Jeacocke (Health and Safety Executive)

The first question was: where does somebody who is asked to move a particular chemical and knows nothing about it get the information on that chemical? Echoing Mr Lilly, if one was really in that sort of situation one ought to go to a professional body, a professional adviser, to get that sort of information.

56. Who ensures that the load which is to be carried is compatible with what was in the vehicle or road tanker before? This should be the responsibility of whoever is moving the load. It is he who knows what was in the tanker before, and I suggest it is up to him to ensure that anything that he is being asked to carry on the second occasion is compatible with what was carried on the previous occasion, if he did not intend to clean out the tank. If he does not know, he should find out. If he cannot find out, I suggest that he should not accept the load.

57. The third question was: who would check the history of the tank or carrying compartment? If anybody is in the business of conveying hazardous substances in a second hand tank without assuring himself of its history or getting an independent assessment of its condition, he ought not to be in business.

58. Mr Lewin then asked whether there was an audit on accidents. There are some statutory requirements about the reporting of accidents during the convey-

ance of certain hazardous materials, but in general they only relate to loss of life or personal injury. The sort of accidents that we have been considering — roll-over of tankers, for example, where there is no loss of life and no personal injury, are not statutorily reportable. News of these comes through the emergency services, and once again this depends on just how severe they themselves consider the incident, or how interested they think the HSE should be in the incident. The pattern is very variable.

59. How is such information used in dealing with proposals for future legislation? That depends on how far one wants to go. In relation to roll-overs and jacknifing of articulated lorries, why not ban them from carrying hazardous loads? Drivers are trained to drive articulated lorries so that they do not overturn, but this happens. The vehicles cannot be banned, so should one put baffles in tanks to stop surge or to put the tank on a low loader type of vehicle as has been suggested. Information from accident statistics must be used sensibly.

References
1. CREMER and WARNER. *Cheshire chemical and allied industry survey.* Report to Cheshire County Council, 1975.
2. LPGITA (UK). *Fires and other occurrences involving bulk quantities of liquefied petroleum gas.* Document 1/77. Liquefied Petroleum Gas Industry Technical Association (UK), Shepperton, 1977.
3. LPGITA (UK). *Liquefied petroleum gas: a guide to handling emergencies.* LPGITA (UK), Shepperton, 1978.

7. Emergency response systems

A. H. SMITH, MCIT, Chemical Industries Association Ltd.

The number of serious chemical incidents is low, but the Chemical Industries Association (CIA) considers it essential that its members should be able to respond to calls for assistance from the emergency services. Communications are basic to any company's ability to respond. This starts with the essential emergency telephone number on packages and vehicles and extends to the availability of a full list of key personnel, their locations plus telephone numbers at work and home. Improved marking of vehicles, part of which is soon to be mandatory, will not reduce the need for companies to be able to respond to calls for advice from the emergency services. CIA is working for the Chemsafe response system to extend in the future to all distribution functions—road, rail, ports, airports and transit warehouses. This will give greater response coverage to assist the emergency services.

The Chemical Industries Association (CIA) considers that any company causing hazardous goods to be conveyed must satisfy itself that suitable arrangements are made for their safe movement and also for minimizing the damage or injury that could arise should an incident occur during distribution. Companies must, therefore, provide information about the potential hazards of products being transported and must perform a supporting role in dealing with emergencies when requested to do so by the authorities.

2. The CIA manual *Road transport of hazardous chemicals* describes good practice which companies should adopt to ensure high standards of safety. The CIA Chemsafe manual *The Chemical Industry scheme for assistance in freight emergencies* is concerned with the handling of an incident should it occur.

3. The Chemsafe manual describes a standard procedure and a long-stop procedure which should be observed by both industry and the authorities. It also gives advice on the arrangements that should be made within companies to ensure compliance with these procedures. This advice is based on tried and proved practices. Companies have found that if the initial work is undertaken systematically the results are rewarding in human terms and in loss prevention.

4. The public emergency authorities (Police, Fire and Ambulance services) and other authorities (British Rail, ports and airports), have the primary role of dealing with all transport emergencies, including those arising from the distribution of hazardous goods. In the case of chemical products the chemical industry has a vital supporting role to perform when requested by the authorities.

Transport of Hazardous Materials. ICE, London.

5. When an incident occurs the authorities need to have immediate information on the possible hazards of a substance and on the action to be taken or avoided. For this purpose there must be:

  (a) effective markings;
  (b) readily available transport emergency cards (Tremcards) or similar instructions in writing;
  (c) more detailed advice or help available quickly when the authorities request it.

6. The provision of this advice or help constitutes the supporting role of the industry. It is therefore essential that manufacturers/traders have available a continuously manned telephone and display the telephone number on the vehicle and package. Industry's response to such requests from the authorities may be the provision of:

  (a) technical advice by telephone
  (b) technical advice at the scene, or
  (c) technical advice and such other assistance at the scene as may be appropriate and available.

7. Chemsafe (Chemical Industry Scheme for Assistance in Freight Emergencies) describes the role of the authorities and of industry in its support of the authorities. Chemsafe operates through two procedures, the standard procedure and the long-stop procedure.

8. The CIA considers that it is a fundamental principle that any company, of whatever size, which manufactures or trades in potentially hazardous goods has an obligation to satisfy itself that suitable round-the-clock arrangements are made not only for their safe movement, but also for minimizing the damage or injury which could arise should an incident occur in any stage of distribution e.g. in road or rail transit, while parked, at docks, at airports, in intermediate storage etc. It therefore follows that all manufacturers/traders concerned should have a suitable procedure for responding to requests at any time from the authorities for technical advice and/or assistance. Additionally, a chemical firm which receives hazardous substances should be prepared to help if an incident is more accessible to that firm's work than to the sender's.

Communications

9. The first requirement is to establish a reliable channel of communication by means of which the authorities can make contact with the manufacturer/trader round-the-clock. This will normally be by telephone. The allocated num-

ber must be manned by personnel able to give or quickly to obtain advice on the best way to deal with an incident.

10. All too frequently in the past the public emergency authorities have reported that calls for assistance have been unanswered because the company concerned only operates in normal day hours or, at best, has installed an automatic message recording device. Where a member firm cannot provide a 24 h manned telephone service at its own premises, other effective communication arrangements should be made so that the authorities can invoke the company procedure at any hour. Possible methods of making such provision include:

(a) entering into an arrangement with another manufacturer (or a consortium of manufacturers) already operating a round-the-clock emergency scheme (the CIA can assist members to join or set up consortium arrangements);

(b) arranging for the automatic transfer of all out-of-hours telephone calls to an alternative fully manned number as for a night line, e.g. to a duty manager;

(c) using an automatic telephone answering device for out-of-hours calls which will provide callers with a fully operational emergency number to call;

(d) using the services of one of the communications firms who would be supplied with a list of the company personnel to be contacted (this list will need to be regularly updated).

11. If a company has a number of locations it is unlikely that they will all be able to provide the full range of assistance to the authorities. If this is the case, then all other locations, offices, and sites should be fully conversant with the company procedure so that they can refer any requests which may be directed to them to the appropriate company emergency centre.

12. Those manning the allocated telephone will vary from a gateman, night patrolman, or factory guard, to personnel in a plant control room or fully operational emergency control centre. In every case the first action must be to obtain and record details of the incident, using a check list. It is recommended that supplies of check lists should be kept by all persons likely to receive emergency telephone calls. The receiver of the call must at all times have ready access to a person who can provide advice on immediate and subsequent action so that this can be conveyed promptly to the caller. Unless the personnel manning the telephone are trained and competent to give advice, their role should be limited to recording information about the incident and operating the in-company communications. They should be specially instructed in these duties and given refresher training periodically.

13. Information on all the company products and materials, their synonyms or code names, their hazards, immediate action, and special precautions, should be compiled and be readily available to all staff who may be called upon to give advice.

14. There must be available at the company control up-to-date lists of competent persons who can be contacted in respect of the different products consigned or received by the company, and it may be desirable to operate a rota system for those on call. Arrangements should be made so that the competent person, can, if requested by the authorities or if he considers it desirable, go to the scene of the incident without delay. In some circumstances the Police may provide transport, in which case the use of their radio communications link allows technical information to be transmitted to those at the scene of the incident while the person is on his way. Experience has shown that the presence of technical support at the scene is greatly appreciated by the public emergency authorities. Contact should be maintained between the emergency centre and those at the scene of an incident until the operation is successfully completed.

15. Some incidents may require the use of special equipment or neutralizing agents. Consideration should be given to ensuring their ready availability and the means of carrying them, and the appropriate personnel, to the scene.

16. Companies should prepare for their own use a full report of each incident, documenting the action taken and the outcome, and making special note of delays or failures in any part of the procedure. Such reports have proved to be a useful way of assessing the effectiveness of the company's arrangements.

Marking

17. A company procedure is ineffective unless high standards of marking are maintained. Detailed hazard marking requirements and recommendations for hazard identification are given in the CIA manual *Road transport of hazardous chemicals* and in *Hazard identification*, which describes the voluntary tanker marking scheme launched in July 1975. It should be noted in particular that bulk road and rail vehicles (if not covered by the voluntary tanker marking scheme), freight containers, packages and Tremcards (or similar instructions in writing), should be marked with:

   (*a*) the product name (as defined below);
   (*b*) the relevant telephone number giving 24 h coverage.

The CIA has recently also published *A guide to the regulations for the labelling and marking of containers and vehicles*.

18. The product name to be shown should be the British Standard or familiar chemical name, or in the case of mixtures or solutions the British Standard or familiar name of the principal hazardous constituents. Additionally a trade name or selling name and UN number may be shown. The name of the telephone exchange as well as the STD number should be shown. On bulk vehicles not covered by the voluntary tanker marking scheme the telephone number should be conspicuously displayed on the rear and both sides of the vehicle. All chemical firms and others concerned in the movement of chemicals should pay constant attention to this aspect of marking, and in the case of inwards traffic, including imports, should require their suppliers to mark vehicles or packages with at least the name of the product in English.

19. In the case of chemicals considered to have a low degree of hazard, knowledge of this is important to the authorities dealing with a transport incident. For such chemicals, therefore, the same indications as for hazardous products are required, viz. the product name, and the relevant telephone number giving 24 h coverage. If, however, it is not practicable to show such a telephone number, the goods should be accompanied by written instructions on their properties and any relevant precautions to be taken by emergency personnel.

20. In drawing up the overall procedure it is recommended that close liaison is maintained with representatives of the local Police, Fire and Ambulance services and that in co-operation with them, exercises are carried out both as part of the training of company personnel and as a means of disclosing any weakness in the procedure.

Procedures

21. Company emergency procedures are the backbone of Chemsafe, but no company procedure can provide prompt and adequate help in all circumstances, unless there is co-operation between firms.

Chemsafe mutual aid arrangement

22. CIA members who are participants in the mutual aid arrangement pool information about their freight emergency facilities—locations, telephone numbers, and range of products for which expertise is available. They undertake to respond favourably to requests from other participants for assistance at the scene of an incident, provided that this is practicable and within their own range of expertise and capability.

23. The principal objective is to achieve better geographical cover, and speedier effective response to requests for attendance and help at the scene of an incident.

Local consortium arrangements

24. These are local arrangements involving several companies under which a single emergency centre serves all members of the consortium. The principal objectives are to avoid the need for each company to establish a round-the-clock emergency centre and to make available the wider expertise of the consortium as a whole. Each firm remains responsible for dealing with its own incidents, but can call upon the facilities of the consortium centre.

Long-stop procedure

25. The further development of company procedures and of inter-company arrangements, reinforced, it is hoped, by those of other industries, should result in the great majority of incidents being dealt with under the standard procedure. It has to be recognized, however, that the circumstances of an incident may be such that the emergency information provided on the vehicles or packages cannot be read. For this reason a long-stop procedure is needed whereby the authorities can obtain advice and/or assistance, even though the manufacturer/trader involved is either unknown or unobtainable, or the product itself is unidentified.

26. The Chemsafe long-stop procedure involves a Chemsafe Centre operated in collaboration with the National Chemical Emergencies Centre at Harwell. The latter has been established by the Department of the Environment as part of the Toxic and Hazardous Materials Group of the Atomic Energy Research Establishment. The centre has a continuously manned emergency telephone through which the public emergency authorities can ask for technical advice in the case of any freight emergency in which the nature of the chemical hazards cannot be ascertained from other sources (e.g. manufacturer, trader or Tremcard). Qualified staff with practical experience over a broad range of chemicals are available on a call-out rota, and the centre's chemical data bank is being continuously expanded with product information supplied from CIA member firms.

27. Often the Chemsafe Centre will be able to give adequate advice to the authorities by telephone. If, however, attendance at the scene is deemed necessary, qualified staff will turn out either from the centre itself or at the centre's request from a more conveniently located chemical company. Accordingly, certain emergency locations of member firms have undertaken to provide technical advice and/or assistance at the scene of a long-stop emergency, if requested by Chemsafe Centre on behalf of the authorities. These locations do not have an overall role within specified geographical boundaries and requests by the centre for attendance at an incident are placed according to a location's expertise and capability as much as to its distance from the scene. These locations do, how-

ever, have an extra commitment and may occasionally be asked to attend an incident involving a product which is not identified or in which they may not have expertise. In order to carry out their role efficiently these locations need to maintain close liaison with other firms in their vicinity and with the local public emergency authorities in order to build up knowledge about the availability of equipment such as neutralizing agents, breathing apparatus and pumps.

28. In a long-stop incident, as soon as the identity of the manufacturer/trader becomes known, he will be expected to take charge of the provision of advice and/or assistance. The Chemsafe Centre and any other location involved should not, however, stand down until it is clear that the public emergency authorities and the manufacturer/trader no longer need their help.

29. The CIA expects that in the long-stop situations, i.e. when the manufacturer/trader cannot be contacted, and particularly when the product is unidentified, the public emergency authorities will telephone the Chemsafe Centre for advice and/or assistance. It recognizes, however, that these authorities always have the option of approaching a local firm direct if they judge that this will result in the required help being obtained more quickly.

30. The CIA keeps in close touch with various bodies responsible for transport services, particularly British Railways, the Road Haulage Association, the British Ports Association, British Airways and the airports. The Road Haulage Association (RHA) has set up a scheme in conjunction with Chemsafe, under which RHA members will, except where special tanks or special loading/discharging systems are necessary, be able to offer physical assistance to the emergency authorities by supplying a suitable empty tanker and competent personnel for the transfer of a bulk chemical load where this is deemed necessary, and safe. Certain RHA member companies will set up regional emergency centres from where the transfer activities will be controlled. The centres will have information on the availability of suitable vehicles and equipment near to the scene of an incident and will arrange for the transporting of tanker, emergency kit and competent personnel to the scene as quickly as possible after receiving a request from the authorities.

31. The emergency team will carry out the transfer operation on request from the emergency services. The latter should already have been advised on the safety of such an operation by the chemical manufacturer's representative, called out under the Chemsafe scheme, but it will still be imperative for the RHA emergency team to satisfy itself that the agreement of the chemical manufacturers or traders has been obtained before transfer is started.

32. Close collaboration also exists with many trade associations concerned with potentially hazardous products.

33. The emergency services have been supplied with an inventory of the

availability throughout the UK of stocks of neutralizing agents for dealing with spillages.

Organization of company team

34. So far this Paper has been concerned with giving a picture of the CIA Chemsafe schemes to deal with responses to call for assistance in dealing with incidents. The remaining part will be of·a general nature dealing with the organizing of a company emergency team and, points arising out of practical operating experience.

35. The selection of a team to cope with emergency calls needs some thought. It should not be just a question of selecting all technical personnel who have knowledge of the products but a mixture of personnel from different functions who have some particular expertise to offer. A team, therefore, could comprise chemists, chemical engineers, packaging and distribution personnel, production management, safety officers etc.

36. There must be sufficient members of the team to allow for absence of personnel on other company work, sickness, holidays etc. The total number in any team obviously has to be relevant to the product distribution pattern. Large tonnage movements from a manufacturing site would probably require from 6 to 10 team members.

37. Personnel in the emergency team as mentioned earlier will need to have available to them the following information in duplicate (one set to be available at their homes and the other at place of work):

  (a) full data on products: hazards, protective clothing requirements, action in case of fire, spillage, first aid treatment, any special medical treatment;

  (b) details of back-up service available from base site and other company sites: names of other technical specialists not in team; availability of vehicles and types, e.g. platform, tippers, tankers; home and work telephone numbers of key personnel (site management, engineering, distribution);

  (c) availability of stocks of neutralizing agents.

38. The team must have transport available to travel to any incident as and when required. In many cases the need to convey a member to give advice is all that is required in the way of transport. However, it is recommended that where there is large movement of hazardous products a van, Landrover or similar vehicle be promptly available to the team and that this be permanently loaded with essential basic equipment: full protective clothing for two people, hand

tools, drum handling equipment, empty packages, small transfer pumps and other more specialized items as are considered necessary, e.g. breathing apparatus, tanker transfer pump.

39. Quick communication is a vital element in dealing with emergency calls. Companies should seriously consider the installation of special emergency telephone extensions for all team members, and an instruction to switchboard personnel to give priority to all emergency calls. The emergency extension numbers must not be made available to other personnel. The availability of team personnel should be known to switchboard operators and, out of hours, to security or other personnel who will receive calls. A weekly list needs to be provided giving this information.

40. Contact between team members going to an incident and those back at base can be valuable, and the fitting of air call equipment in the emergency vehicle will provide this facility.

How the team operates

41. The team is selected, the data on products are available, full information on back-up services has been provided and the communications systems have been checked so that in theory the team can cope with emergency calls.

42. It is not possible to anticipate the nature of emergencies which may arise in the future. The team leader can, however, regularly call his team together to go through hypothetical incidents to decide what advice or action would be required to assist the emergency services in such an eventuality.

43. It is important at this juncture to point out that the emergency services are responsible for dealing with incidents, and there is no question of a company emergency team taking over that responsibility. The requirement from a team will be to give accurate technical advice as promptly as possible by telephone and to proceed to the incident to give further assistance if requested or if the company considers it advisable in the general public interest to do so. The provision of vehicles, neutralizing agents, special pumps etc. will be part of the assistance offered to or requested by the emergency services as a follow up to the technical advice.

44. The circumstances surrounding a particular incident will have some bearing on the advice given. This does not mean that the basic hazards of the chemical alter e.g. toxic, corrosive, flammable etc., but that factors such as location or type of accident may require additional advice. The presence of toxic fumes is a more difficult problem in built-up locations than in open countryside. Spillage of flammable products is again a greater hazard in the former location. The type of accident affects the advice and any requirement to attend the scene.

A multi vehicle collision will result in there being a mixture of chemicals and other substances which could increase the hazard substantially.

45. The vital ingredient towards giving an accurate response to the emergency services is to complete the check list accurately when receiving the first emergency call. The name of the chemical should be spelled out to ensure that there is no doubt e.g. receipt of telephone messages about incidents involving hydrochloric or alternatively hydrofluoric acid could lead to wrong or insufficient advice if care is not taken to check the actual product involved. An additional safeguard is to have readily available in all traffic departments details of consignments sent out over the past 48 h so that confirmation of the load can be obtained quickly.

46. The introduction of the Hazchem tanker marking scheme referred to earlier has enabled the emergency services to take immediate action to deal with incidents involving bulk vehicles, but the Chemsafe response system is still essential to cover additional hazards arising out of particular circumstances or extra general information. This is, of course, the reason why the emergency telephone number is part of the composite sign.

47. It is useful to stress again that, following any incident in which the Chemsafe response has been requested, the emergency team personnel involved submit a report promptly. Two important points are first, the cause of the incident (and any recommendations to be made towards avoiding future incidents) and second, an assessment of the team's response.

48. The Chemsafe scheme applies to road and rail transport and in the future is likely to figure in the emergency plans for ports and airports dealing with freight movement. The application of Chemsafe response on a company basis will obviously be available to shipping companies in connection with incidents on vessels, but one has to accept that there may be problems arising from the carriage of a variety of materials in large tonnages where the advice for dealing with a single product may conflict with that for others stowed in the vicinity. There will, therefore, have to be co-ordination to decide the action to be taken in such emergencies.

49. The CIA is rightly putting much effort into the safe carriage of chemical products, but this is no justification for relaxing the Chemsafe response system. Finally, it is important to point out that the number of serious incidents arising out of the movement of chemicals by road and rail is very low, particularly so in relation to the tonnage movement.

Discussion: Paper 7

Mr M. Grylls (Member of Parliament)
It is only too easy to produce laws and regulations in the field of health and safety with which everyone agrees, but it is industry, local government and civil engineers who have to understand them, interpret them and work with them, and, of course it is industry which has to make a profit with them. The more we know about the reaction of industry and the professions to laws and regulations, and the greater the flow between us, the better.

2. In travelling around the country I find that people are increasingly aware that the frontiers of knowledge are moving ahead extremely rapidly, and those living near plants and on busy roads feel that they are living in dangerous places. Most big industrial and chemical companies who have got in touch with the people who live near their factories and plants and have told them in simple terms what is going on—that every industry has some hazards but that they are being as efficient as possible to minimize these—get a very good flow-back from the public. The feeling that people outside the factory gates should not be told what is going on is really out of date and should be discarded.

3. The interface between the public and its representatives and the industries which have to deal with many of these nasty and difficult products is something that must be improved, I think. But industry and the professions must make themselves as thoroughly professional as they can. That professionalism will encourage the public to believe that things are well run. Industry must have well thought out, and perhaps above all, practised and tried emergency plans. This applies just as much to the site as it does when materials are moved on the roads.

4. The Chemical Emergency Centre at Harwell is doing fine work. There is much the legislature can learn from a place like Harwell and from the Health and Safety Executive, who meet industry eyeball to eyeball and talk the same sort of language.

5. As far as transport is concerned, as a politician I judge that the public worries a great deal about this. If any incident occurs, the television cameras are instantly there. So everyone assumes that every tanker going along the roads is carrying nasty chemicals. The fact that the majority are carrying harmless substances like milk or even wine does not occur to them. It is important to let people know that matters are well organized.

Mr S. C. Baker (HM Inspector of Fire Services, Home Office)
The fire services consider that the chemical, petrochemical and petroleum industries, the hauliers and the waste disposal contractors, are to be congratulated

on the development of the Chemsafe scheme and other voluntary operations towards safety in the transport situation. These schemes are welcomed by the firemen.

7. The main criticism seems to be the 3h wait for somebody to come and help. This is a long time. The Home Office is making a survey of incidents involving dangerous goods, and from the evidence the response is unsatisfactory in only 2% of cases. Were there no Chemsafe arrangements, the waiting time would not be a matter of hours: it would often be to infinity. We would get nobody.

8. The proliferation of different labelling will obviously cause confusion. There is already a United Nations system of labelling for transport and a new Council of Europe system for labelling packages and other goods, the orange label. This directive was published in 1967, before the UK joined the Common Market, and therefore I have no doubt that it will have to be implemented, but I make a plea to all those involved in law making and regulation making, that we must strive for total long term *international* harmonization. This will help the shippers and everybody in the industry concerned with the movement of dangerous goods.

9. Paper 7 refers to the proposed marking of buildings. This is not going to be a simple thing to achieve. In nearly every case there is a mixed load, and it will be very difficult to give any meaningful information to firemen and others who have to deal with incidents in buildings involving dangerous goods. In any resolution of problems, the scheme must be kept in the form that is familiar, though not necessarily exactly taking over the Hazchem scheme. There are many problems to be overcome before a proper system is evolved, but the emergency services would like to see a format of the kind that we know now for road and rail, and to this end, of course, industry's co-operation will be absolutely essential.

Dr F. S. Feates (Harwell Laboratory)
I should like to comment on this problem of the speed of response. It is clearly gratifying that it is only in 2% of cases that speed is not adequate, but we should not be complacent.

11. I would like to ask Mr Smith if he thinks the Chemical Emergency Centre has got it right, or if there is something else that we can do? The mechanism to deal with these situations is really two-stage. If it is possible to identify the chemical manufacturer or supplier who has the particular specialist knowledge and he can get to the scene of the incident quickly, this is fine. However, if the incident is in Southampton and the manufacturer is in York, even if a helicopter is used it would take a couple of hours, and if he is in New York, it is even more difficult. What do you do then?

12. Under the present system there are two sorts of specialist: the local

specialist who knows something about accidents and chemistry, and the technical specialist who knows a lot about the product, and they are reached by telephone. My feeling is that this part of Chemsafe is not working as well as it might. We are not getting a local man to the scene to advise on the spot and getting him advised in turn by someone with specialist knowledge of the product. Does Mr Smith think that this should be working better?

Mr Smith (Paper 7)
In reply to Dr Feates, I did say that the most important thing in cases of dealing with an emergency is to get the name of the manufacturer, and if that is available to go direct. There have not been many cases of delay, where the incident is far removed from the location of the company's emergency centre, and where the alternative use of a third party as part of a mutual aid scheme may not be possible. The reason why delays have occurred, should be established; whether it was because the mutual aid scheme was not working efficiently or whether the delay was due to the provision of special equipment which required time to get to the job.

14. The question raised by Mr Baker about labelling of buildings and tanks has been discussed a fair amount recently, and I think people are coming round to the thinking that he expressed, that the use of similar labels to the Hazchem type on a building containing a whole variety of products is not a good idea. Although these labels could possibly be used for a single large storage tank where there is no question of a variety of chemicals being involved. It would be a good idea, as he suggested, to use the same orange background, to signify that it was a warning notice and an indication of hazards. What is required is an International system.

8. General principles of risk assessment

F. R. FARMER, OBE, BA, FInstP, HonFSE, Safety Adviser to the UKAEA, Visiting Professor, Imperial College, London

It has been common practice to assess the risk of hurt from accidents by extrapolating past experience in similar situations—as in buildings, bridges, dams and specifically in road, rail and air transport. This type of risk assessment may give a reasonable picture of that which is most likely to happen. Is that the information required when considering potentially hazardous situations, or should the result of some defined 'worst' situation, as of impact at 60 miles/h or more, followed by fire, be used? Experience in risk assessment in the nuclear field indicates clearly that the consequence of accidents cannot be adequately described by a 'most likely' or by an artifical 'worst' situation. In general there exists a range of possible end results some more likely than others. An awareness of this spectrum of risk and consequence is an important factor in selecting designs and operational procedures to minimize risk in areas in which the hazard is potentially very high.

There is a growing concern about the possibility of major or unusual hazards, accelerated perhaps by the accidents at Flixborough and Seveso. Accidents are widely reported and in some cases, particularly if there is an association with radioactivity, even minor events may assume major significance. It is important to take note of this concern; there is considerable public debate and interaction on proposed industrial activities and there is a significant increase in the size and complexity of these activities.

2. Against this background I wish to explore the assessment of risk and the form of communication of any assessment. Much effort has been and is being applied to the technical assessment but little progress has been made in communication with the public.

Background

3. The chance of accidental death at work or at home is decreasing and this trend should continue. Most accidents involve only a few people and arise from a large variety of causes, but in total lead to the death of about 10 000 people at work or home and a further 7000 from road accidents. These are large numbers and deserve a sustained effort to reduce them; however it is not the purpose of the Paper to explore this area of accidental risks, but rather to turn to the unusual event which may constitute a major hazard or have the potential so to do.

Transport of Hazardous Materials. ICE, London.

4. It is possible to imagine accidents in industrial installations or transport which could cause the death of tens, hundreds or thousands of people. The likelihood of such a catastrophe is small (or is believed to be so) yet the consequence could be very serious indeed. The reaction to an event such as Flixborough is far greater than the combined reactions to the 50 accidental deaths which occur each day in the UK.

What to assess?

5. As a member of the Advisory Committee on Major Hazards of the Health and Safety Commission, I have a particular interest in the identification of major hazards and the assessment of the risk they pose. It is obviously impossible to survey, in depth, all industrial installations or to ask all managements to carry out their own survey with respect to major hazards and report thereon. This would present an impossible load on industry and on any supervisory body. The effort is not available and the attempt to spot the unlikely event would detract from the need to watch for the more likely events.

6. The same applies to a review of transport operations. What are possible ways of narrowing down the selection? A start might be made by assuming that all well-established operations are satisfactory and if subject to steady updating should remain so. Hence there would then be a need to look only at new operations or those recently introduced, or on occasions when substantial changes are proposed, such as in the type of transport, speed, routes etc. But is there not always a belief in the adequacy of existing procedures before and until a serious accident occurs?

7. Another approach would be to look for hazard potential. This is not as simple or as useful as it may appear. In theory, the potential is determined by the quantity of material, its toxicity, flammability or explosive capability. In practice most accidents do not reach their full potential as the consequence depends on many factors: locality, population density, weather conditions, emergency procedures, etc.

8. A good example of the way in which a serious accident fell far short of possible consequence, in terms of fatalities, is the derailment of a train at Illinois in 1970. The train was derailed early in the morning. The train included ten tank cars, each containing 33 000 gal. of propane. The fires and explosions were sequential at intervals of 30 min. or so; there was much damage, estimated at $3 million, but no deaths.

9. From a review of many accidents, it has been estimated that explosive type accidents, on average, kill about one person per ton of explosive or per ton of flammable gas released. Hence the Illinois accident could, under some conditions, have killed perhaps 100 people from one tank car explosion or up to

1000 if all tanks had exploded simultaneously. It is unlikely that all could fail together to give one big bang and it is, of course, the degree of unlikeliness of various combinations of the type of accident, the site, the weather, etc., which are crucial to the assessment of risk.

10. The release of 100 tons of Cl_2 at Baton Rouge, Louisiana, did not kill anybody: a favourable wind helped to blow the cloud of gas across the Mississippi river and away from areas of heaviest population density. Had the wind been blowing towards the population centre, it is still possible that few, if any, casualties would have occurred, as the release was spread over a period of 6 h or so. If one were assessing the probable result of releasing 100 t (based on previous accident histories) the estimate would be some 20-100 fatalities. This estimate depends on the way past evidence is used. The report by Simmons et al.[1] reviewing Cl_2 spill accidents from 1934 to 1974 shows no deaths from railroad accidents but about one person per ton from accidents to storage tanks.

Prediction based on extrapolation of experience

11. Examples of predictive techniques appear in reports assessing the hazards of transporting Cl_2. Westbrook[2] compares road/rail for the UK, Simmons et al.[1] assess potential fatalities for transport in the USA and Lautkaski[3] for transport in Finland. I refer to these papers to illustrate methods rather than to agree or disagree with their conclusion.

12. Westbrook uses a deterministic approach. He uses data on the frequency of accidents, and modifies this number to arrive at severe accidents, i.e. to estimate the fraction which would cause leakage. He then takes a specified flash and leakage rate, and calculates the area affected by normal dispersion in average weather within 24 min., using available data on toxicity of Cl_2. Finally he allows for 'escape'—a speculative but important factor. As a result of this calculation, Westbrook arrives at specific numbers predicting one fatality every 100 years for the current scale of operation at 200 000 tons/year.

13. The other authors used distributional factors in some parts of the assessment, such as weather or population distribution, with ascribed probability that particular conditions may exist at the time of the accident.

14. It is not to be inferred that one type of assessment is right and another is wrong. They are giving different information which should be consistent if applied to the same problem. The possibility of one fatality every 100 years does not enable one to conclude that it is most unlikely that any one person or more might be killed in this decade, or how unlikely it is that an accident might kill 100 people. If this information is needed, the distributional factors must be retained throughout the analysis rather than only the average values being used.

15. The three papers referred to indicate the nature of the information required for typical risk assessment, such as:
  (a) the frequency of accidents in road/rail transport;
  (b) assessment of severity (this might be extended);
  (c) nature of site, e.g. spillage on land or water affecting (i) rate of release of hazardous material; (ii) weather distribution; (iii) population distribution (it may be necessary to relate this with (a) and (b));
  (d) toxicity or likelihood of fire or explosion on site or at some distance away;
  (e) effectiveness of emergency procedure; various models have been used; rather than use one value as in Westbrook (that all hazard terminates at 24 min.) it might be assumed that there is some chance, perhaps 1 in 10, of low effectiveness possibly linked with weather and site.

16. Using some but not all of these factors, two of the reports give a distribution of possible fatalities with ascribed probabilities. Simmons *et al.* for the USA give results similar to those in the Rasmussen report WASH 1400 (possibly the same source was used). For the transport of 2 million tons/year, the chance of Cl_2 spillages causing 100-200 casualties is given as $\approx 10^{-2}$/year, and 800-1000 casualties as $\approx 10^{-3}$/year.

17. In comparison, the report from Finland for the transport of 150 000 tons/year gives the chance of Cl_2 spillage causing 100-200 casualties as $\approx 10^{-3}$/year, and that for 800-1000 casualties as $\approx 10^{-4}$/year. These results are reasonably consistent and do not invalidate Westbrook's conclusion.

What is reasonable risk?

18. There have been a number of meetings or publications concerned with the acceptance of risk. There seems no concensus of opinion on this point. The view seems to be rather that all risk of death or injury should be prevented if reasonably possible; hence the need to establish the adequacy of preventive measures against the risk of hazard materializing.

19. There is no basic guidance on what is considered reasonable in advance of any accident other than the observance of current codes and recommendations of good practice.

20. I suggest at least two factors should be taken into account:
  (a) would the accident (major hazard) have been prevented by further modest effort;
  (b) irrespective of (a), would a risk assessment have shown the chance of the event and its consequence to be markedly higher than the current risk/consequence level in all other activities in the country.

In order to answer (*b*) a better picture of the risk of serious accidents than I believe currently exists is essential. This is true for industrial installations: quoting from the Health and Safety Commission news release of 4 August 1977: 'Preparation of the regulations will enable a complete picture to be built up, for the first time, of the number of potentially hazardous installations in Britain, and lead to the development of tighter controls.'

21. I am not particularly familiar with transport problems. It may be that many risk assessments have been made, although I doubt whether they have covered a range of probabilities *vs* consequences. If not, I suggest the risk of major hazard transport needs to be assessed in order to build up a picture in parallel with that being developed for fixed installations.

22. Any exercise of this type is bound to be selective and progressive; initially a few examples could be chosen of operations judged to be potentially hazardous and subsequent studies developed from their findings.

References

1. WESTBROOK G. W. The bulk distribution of toxic substances: a safety assessment of the carriage of liquid chlorine, *Loss prevention and safety promotion in the process industries*, Elsevier Scientific Publishing Company, 1974.
2. SIMMONS J. A. *et al.* Risk assessment of large spills of toxic materials. *Control of hazardous material spills*, Conference, San Francisco, August 1974.
3. LAUTKASKI R. and MANKAMO T. *Chlorine transportation risk assessment.* Technical Research Centre of Finald, Nuclear Power Engineering Laboratory, Report 27, Helsinki, December 1976.

Discussion: Paper 8

Mr V. C. Marshall (Director of Safety services, University of Bradford)
Mr Farmer has exemplified the fact that prediction is a very tricky business, particularly when it is concerned with the future! The authorities he quoted represent the more sober end of the business: some of the figures have ranged as high as 8000 fatalities for 1 ton of chlorine spilt.

2. The question of mean results is perhaps more important than he has suggested. In any statistical distribution there is a mean and some measure of variance. In this field the curve is always heavily skewed because while there is a mean, the minimum is zero, because whatever the curves might say, there is no probability of actually *creating* life as a result of one of these accidents! Therefore there is a sharp boundary at zero at the low end, but at the high end there is a long tail off, with lower and lower probabilities attached to it.

3. Assessing a mean from historical experience gives some opportunity, however roughly, of making a comparison between the relative importance of hazards to be faced, because to classify these emotionally is to admit that nothing can be done about them. Statistics give some sort of perspective. In a recent paper[1] I suggested that clouds of vapour and high explosive are equally dangerous, tonne for tonne, and that they are likely to kill possibly 1-5 persons per tonne of material. Chlorine on the other hand, although there is a great deal of horror on the part of the public because of its association with warfare, is, in my view, not as dangerous as high explosive. Chlorine has something of the order of 0.4 fatalities per tonne, based on historical experience, though this experience is limited. I have traced only 30 incidents in the literature and I am hoping that more information will be forthcoming. The figure for ammonia is even lower: of the order of 1/40th of a person per tonne released, based on 6 recorded instances.

4. This is a case where the bias is likely to be on the side of pessimism. What gets into the press is the sensational incident—that associated with fatalities—whereas things which do not produce fatalities are often ignored. The Paper quotes the incident at Baton Rouge, with possibly the largest recorded release of chlorine in industrial history (though somewhat smaller than the gas clouds used in World War I), and exactly nobody was killed. If this had killed 40 people, I suggest the incident would be well known, but even among engineers and people concerned with these things there are many who have never heard of it at all, even though it is a recent event.

5. There is a necessity for the companies that have the information to make their information more readily available. If it is really good news as they claim, I see no reason why they should not release it. There is indeed a case for theoretical

studies, but papers should admit, where this is true, that the results do not accord with historical experience. There is a great deal to be learned from theoretical studies, particularly in developing scaling laws. It is quite clear that high explosive becomes less and less deadly per tonne, the greater the size of the accident: 100 times as much high explosive does not kill 100 times more people. This is related to a scaling law, but nothing has ever been published which would give any scaling law for toxic releases.

Mr C. Harris (Imperial Chemical Industries Ltd)
This problem of risk assessment is complex and demanding of resources of time, experience and data. These are not available to the smaller concerns, and it is left largely to the major groups to develop this approach. The Mond Division of ICI has been actively engaged on this for some years, assessing all its plants and operations. Transport is one of the areas covered. The paper by Westbrook quoted by Mr Farmer was based on work which had been done in 1971. At that time the only data available were from American sources. Westbrook developed a method of assessing whether road or rail transport was safer for chlorine, but the data were not precise enough to differentiate.

7. The more important result of that exercise was the detail of how often these incidents were likely to take place, the causes for the loss of containment and the measures which could be taken to reduce them. ICI has slowly been developing a policy of modifying operations, vehicles and containers, so as to reduce a risk, even though at the time we believed the risk to be acceptable. We have learned, for example, that in the case of road tankers one of the vulnerable points in an accident is the valves: if the valves stand proud of the tanks they are more likely to be snapped off or damaged than if they are sunk into the tanks. The move is now towards recessing the valves of such tankers within the outline profile of the tank barrel: they must then be safer because they are protected by more steelwork than they were originally.

8. ICI has also examined the routes along which pass some of the more hazardous traffic. This has been done on a quantitative basis from road accident data and from data on our own vehicles. We have come to very similar conclusions to those of Mr Ashton. The ideas of the local authorities and operatives on Teesside and the quantitative assessment come to the same conclusion: that certain types of road present lower risks than others.

9. Many tankers are now fitted with crash barriers along the side. This is an idea that was started at ICI because analysis showed that a certain proportion of those tankers which were punctured in crashes were unprotected on the sides. With barriers, the chances of the barrel being punctured are reduced, as was confirmed by a recent accident on the M5 when a tanker with crash barriers

fitted overturned and there was no damage of any significance whatsoever to the barrel, although it was believed that the tanker had been travelling at 60 mile/h at the time.

10. The Westbrook assessment is now an old one, and has been since revised using data now available in the UK which are relevant. It does not change the conclusions, so we are pressing on with those improvements which can be made easily and relatively cheaply. The paper by Simmons was done independently by the University of California under contract to one of the American government agencies for the purpose of the Rasmussen Report. An interesting point is that Simmons concluded in his work that the fatalities which had actually occurred in the USA were over an order of magnitude lower than he predicted. There are mitigating factors, some of which he was aware of, but in his study he did not make appropriate allowances. This is one of the difficult areas. What allowances should one make, and how can they be quantified in an assessment?

11. The assessment by Lautkaski and Mankamo was concerned basically with a rail transportation problem. A significant point was that they had more statistics on rail accidents from Finish Railways than are obtainable here from British Railways. Statutory requirement for reporting accidents is laid down in various UK laws. Unfortunately not all rail accidents or dangerous occurences are reportable.

12. What of the future? Modelling must continue. Any significant change in transport patterns will be looked at, using our models, to see what is the best way to try to improve them. Certain parts of the models have to be further validated by tests. Some of the models published in the literature or used by other authorities and organizations have not been validated, and must be validated before they can be accepted. The Americans have examined eight different models for liquified natural gas, for instance. In the case of toxicology, it is very difficult to do any experimental work directly relating to human beings, and this is another area where data are few. Co-operation is essential, because in this field there is no one organization that understands everything and can develop a complete model. Little can be achieved by working in isolation.

13. I have a final point on the historical records. One of the problems is that many data are stored in the files of industry. A great deal of effort is needed to get them out, but this is worthwhile. In the case of chlorine, for instance, the bulk of it has been collected by the chlorine industry, who have imparted confidential status to it; I think we must now see if a digest of that information, which is useful for modelling not only in the UK but worldwide, can be released to further knowledge in this field.

Mr Farmer (Paper 8)

I agree that the further down the road one goes in prediction, the more un-

DISCUSSION: PAPER 8

certain it is. Much of the reason for doing it is not to arrive at an absolute figure or to believe the absolute figure, but to have an exercise between different situations using the same basic assumptions.

15. Mr Marshall and I agree in that we do not like the extremist; the extremist does not help the subject at all. However, I disagree with him in that it is not necessary either to look backwards at history or to look upwards at the heavens. There is a third choice, and that is to use the available data. In going from an accident at one place to an accident at another, data on population distribution, on wind and weather pattern, wind speeds, and many other factors, are required. From these and from the hard work done over the last 10-20 years by such organizations as Harwell and the CEGB, it is possible to glean information which means that in predictive work we are not merely consulting the stars.

16. One recognizes however the uncertainties when one gets far removed from current historical records. Any further prediction must be based on and be consistent with that which history tells us.

17. In relation to Mr Harris' remarks I have checked Simmons paper again and find the last part of his analysis to be unsatisfactory. He develops seven ranges of potential mortalities each with a specified probability (range 10^{-2} to 10^{-4} per year). He then takes a component from each range as a contribution to expected fatalities per year. For this to be valid we would need over 10^4 years without changing the frequency and mortality rate.

18. More correctly his analysis estimates one accident every ten years and any one accident should be the most likely, i.e. one with few casualties. His diagram in Fig. 5 can be read as showing the chance of killing 10 people to be far less than once in ten years. There is then no evidence to discount his consequence/frequency diagram, it may or may not be correct. It was probably high as Simmons remarks, post accident action to control the effects and number of injuries was neglected.

Reference

1. Marshall, V. C. How lethal are explosions and toxic escapes. *Chem. Engr.*, 1977, No. 323, Aug., 573-577.

Summing up

Professor B. H. HARVEY, Visiting Professor, Dept. of Safety and Hygiene, University of Aston: latterly Deputy Director, Health and Safety Executive

Paper 1 looked at the problems of developing legislation in an international framework. The first point made was that nobody has ever said that things can be made absolutely safe. The second was that the Health and Safety Executive cannot make things safe. What it can do is set up guidelines which might enable the people who have the responsibility for these jobs to make them safe for themselves.

Paper 2 covered the Harwell scheme and was essentially descriptive. The discussion raised several points of great interest. There was the question of water and sewage. There was the suggestion that emergency arrangements might make it difficult for other people who have to pick up the bits. It appeared that the local authorities did not feel fully in the scheme, but if the scheme was simplified so that ordinary people could operate it in an emergency it would lack comprehensiveness and conflict might ensue. Two other important problems were mentioned: the first was that of mixed loads to which no adequate solution has appeared; the second the transport of waste.

Papers 3 and 4 dealt with the constructional problems of vehicles. While the best practices might have been followed, I did not feel that this satisfied the conference. Matters came out in discussion such as fatigue problems, stability of loads, standards of fittings, etc. and also the difficulties involved in loading and discharge of tankers. One astonishing outcome of the discussion was that half the speakers thought there was too much regulation and the others that there was not enough. On the training of drivers, the dangers of isolating a driver from the gravity effects of his load were pointed out. Has the point been reached where handling a load of this kind may be an activity requiring instrumentation for the driver? There was no mention of research into vehicle design, which I feel is a pity. I would also ask if anybody has done work on energy absorption devices so that plastic containers might deform without disrupting on impact?

I had sympathy with Mr Ashton (Paper 5) who pointed out that with the will the Cleveland scheme could be extended throughout the UK. The discussion confirmed that this essentially required a much broader concept of planning in relation to the handling of dangerous materials, whether in static or in mobile situations, and indeed, in pipelines.

SUMMING UP

The need for harmonization of emergency procedures came out strongly in the discussion on Paper 7. It was also seen that there were problems in back-up in one way and another with various companies who might be involved in the schemes. This is an area to which greater attention must be given. Much has been done, but I believe there is a good deal more that could be done.

Risk, covered in Paper 8, is something which I believe requires far more analysis. Paper 1 suggested that the public have equivocal views about risk. Even if their views are without any logical basis, I suspect that they are entitled to them. If they take more notice of 50 people being killed at one time than of 50 people being killed *seriatim*, that must be taken into consideration. What energizes the public's concern is the feeling that the risk is being imposed upon them, that the tanker is an unknown risk. It is not necessarily serving their interests, although in the long term it may be improving their standard of life. This is a reasonable attitude and I would like to see much more work done on risk in the future. The pooling of information and the waiving of confidentiality concerning things which have gone wrong would be of great value.

In the last 300 years many historical events—wars, revolutions, social and political upheavals—have hit the headlines. But the progress of technology has gone inexorably on, almost untouched by these events. It is like an accelerated evolution which will inevitably continue. Technology must be considered as an area in which vital decisions must be made. Humanity cannot afford the luxury of allowing technology to continue uncontrolled.

Transport is one aspect. In 1900, coal trains went from the collieries to the towns, the seaports, etc., and the transport of dangerous materials, so far as there was any, was confined to the rails. There was no transport on the roads. In the past a factory contained most of its dangers within its own confines. Now raw materials of all kinds are on the roads and out in the community. One of the effects of technology is that it penetrates further and further into the home, and it must be assumed that this will continue. Transport will therefore be more penetrative of society.

I believe that the first thing to be done is to find out what is going on, who is doing what, what information there is in this field. That information should be pooled. Who better to act as a clearing house for information than the bodies organizing this symposium?

Where risk is concerned, it is vital to find out whether we are dealing with folk lore or facts. There are a great number of statistics on accidents on the roads, but the near misses never get into the statistics. Therefore the statistics are weighted on the side of disaster, on a spectrum with nothing at one end and total catastrophe at the other. A more valid distribution might come out of a pooling of resources.

SUMMING UP

There may need to be further research into risk itself, to see what is meant by the term. But let us not forget that when the ordinary man takes his car voluntarily out on the road he should not be at more risk than if he voluntarily stays at home and risks falling downstairs.

Legislation is the next area, but the legislation must be comprehensive and I suspect that the problem of harmonization with the United Nations, the EEC, IATA and all the other bodies is difficult and time-consuming. I would advise regulations which are far more flexible. The argument between the proponents of less legislation and those who want more may simply mean that the legislation is of the wrong kind. More flexible legislation would allow technical developments to take place within the framework of that legislation, perhaps supported by codes of practice. The Robens concept of self-regulation requires that those in industry doing the regulating are competent to do it. This may involve the whole question of competence and incompetence in terms of the engineering profession as a whole. Indeed, if anyone is to do this, surely the engineering institutions are the right people.

On design, what is the ideal vehicle? There must be something that satisfies a reasonable number of options. Research is probably needed. I believe that the concept of energy absorption, etc. might be looked at.

Identification and emergency are two areas which are already on the right lines. I would like to see the Harwell data bank considerably extended. I believe that it would be better if it contained medical information which might be of value in other respects.

Lastly I think we need a think tank to look at where we are going. We have let technology roam free and it is time to pull in the leash and make sure that it is going where we want it to go. There has been little mention of rail transport. Information suggests that road and rail are perhaps equally balanced in terms of risk, according to Mr Harris. The fact that the British railway system has been virtually truncated does not mean that it would not be possible to move large and dangerous loads on lines which do not in fact go through large centres of population. There is still in the UK a network of railway lines which are not much used. Could there be a co-ordinated system of road and rail, whereby some things could go by rail and others by road? This could not be done unless it is examined as a specific issue.

Pipelines were mentioned only once. In theory the pipeline is probably the safest of all forms of transport and is used for many products. A co-ordinated look at these problems, considering what should go on rail, what by road and what by pipeline, might come up with a scheme which would achieve the maximum commercial and safety success. This cannot be done on the basis of deciding to think about it when the pressures demand a decision. As a result of technological advance dangerous materials are going to penetrate further and further into the domestic scene. What is needed is someone or some organization to consider how we should cope with this as a nation and as a society. Do we not need long-term research into what is best? It has always been my belief that there is need for some philosophy in dealing with these problems. It is all too easy to become immersed in day-to-day matters. I would ask that we take a broad overall view. There is a case for occasionally postulating an ideal world and considering what we would do in that ideal world.

I believe that we have reached a point where we have to decide whether technology should be our master or our servant. I believe that a crisis of opinion is rapidly approaching and what we have been discussing today has something to do with it.